DIAGRAMS

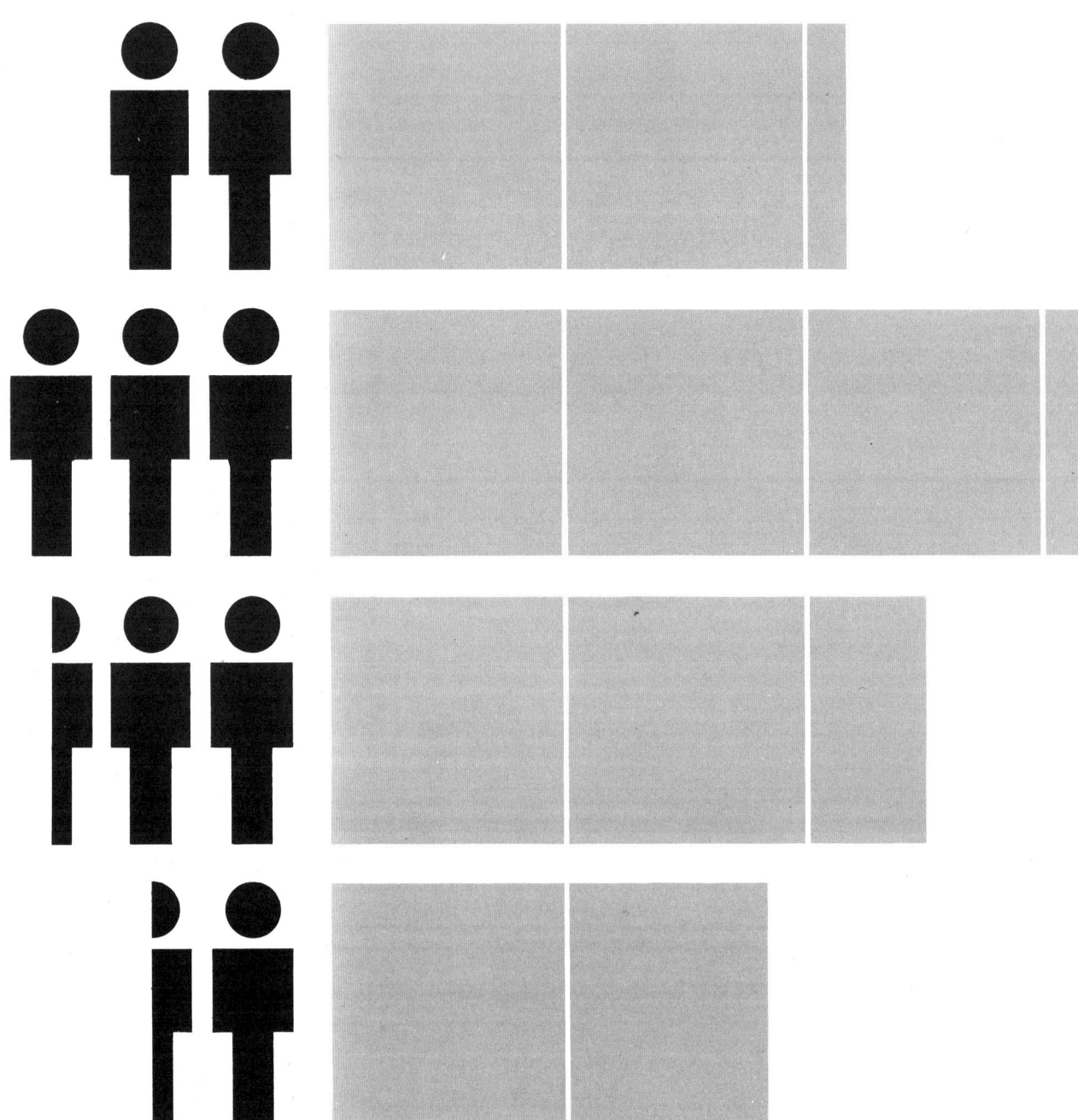

ARTHUR LOCKWOOD **DIAGRAMS**

A VISUAL SURVEY OF GRAPHS, MAPS, CHARTS
AND DIAGRAMS FOR THE GRAPHIC DESIGNER

STUDIO VISTA LONDON
WATSON-GUPTILL NEW YORK

Designed by Arthur Lockwood
© 1969 Arthur Lockwood
First published in London 1969 by Studio Vista Limited
Blue Star House, Highgate Hill, London, N.19
Published in New York 1969 by Watson-Guptill
Publications, 165 West 46th Street, New York 10036
Library of Congress Catalog Card Number 77-82136
Made and printed in the Netherlands
by NV Grafische Industrie Haarlem
British SBN 289 37030 2

Contents

6 Introduction

1 Statistical diagrams

7 Line graph
18 Divided (compound) line graph
20 Logarithmic graph
22 Scatter graph
26 Bar (column) graph
28 Divided (compound) bar graph
34 Floating bar graph
36 Population pyramid
40 Bar diagram
42 Block diagram
46 Divided rectangle
50 Circular graph
52 Divided circle (pie or cake graph)
56 Pictorial graph

2 Explanatory and statistical maps

69 Weather map
72 Route map
76 Games and battles
78 Colour
80 Texture and shading
82 Line and tone
84 Bars
86 Blocks
88 Squares
90 Dots and circles
92 Circles and divided circles
94 Symbols – non-quantitative
96 Symbols – quantitative
98 Distorted map (cartogram)
102 Ray map
104 Flow line map

3 Explanatory diagrams

110 Flow chart
116 Technical
118 Organization chart
122 Botanical/medical
124 Family tree
128 Tabulation
130 Time chart

Technique

140 Three dimensions
142 Engineering drawing

143 Bibliography
143 Acknowledgments
144 Index

Introduction

The aim of this book is to make the graphic designer more aware of diagrams, their use and design.

The book has three parts, each covering different aspects of the subject: statistical diagrams, explanatory and statistical maps, explanatory diagrams. There is a final section which briefly discusses problems of technique and includes notes on books that deal in more detail with certain aspects of the subject.

The various ways of showing statistics or providing explanatory drawings by diagram or on map are described. The best uses for and the disadvantages of each method are discussed. The text is fully supported by the illustrations and their captions. There is as wide a selection as possible with examples from many countries: the United States, Germany, France, Sweden, Japan and other countries. Diagram methods are part of the language of international communication. The diagrams come from a wide variety of sources: atlas, encyclopedia, school and university text book, scientific magazine, newspaper, advertisement, and television. It is most important to remember their origin, since this influences the method used, the amount of information given and the kind of labelling.

The illustrations in this book are important. The designer will see more quickly from example than from description the effectiveness of using a particular kind of diagram. But not all illustrations are equally successful. The diagram produced by the statistician, geographer or sociologist can convey a great deal of information in an interesting way but may lack the graphic distinction which a designer would give. And, conversely, the diagram produced by the designer may have all the current graphic clichés but fail in the aim of getting over information clearly and accurately.

The illustrations come mainly from recent material. The caption for the diagram gives the source or origin, whether book, newspaper etc. Where necessary, it states whether the diagram was in colour. It has been impossible to reproduce every diagram as originally shown, and some of the effectiveness may be lost by reproducing in monochrome. If the diagram has been reduced excessively, this is also stated, since size affects the appearance, especially where labelling is concerned. There is a brief description of what the diagram is showing, and, where possible, if similar information is shown elsewhere in the book in another form, cross references. I have sometimes included a brief comment on the effectiveness or otherwise of the diagram to supplement information in the text.

There is no attempt to give this book a historical perspective, although some of the pioneers in the effective presentation of information in visual form are included. The pioneers include magazines such as *Fortune* and *Scientific American;* firms such as the Container Corporation of America, and particularly the Isotype Institute, London, whose approach and achievement has affected all designers of diagrams even though they may not be aware of their debt.

Firms who understand the diagram problem and effectively use them today are still relatively few. *Scientific American* continues to maintain its lead in imaginative presentation of scientific information. Also in the magazine field, the *History of the 20th Century* is exploring exciting ways of presenting statistical information. The *Reader's Digest* has broken down some of the conventional approaches of the atlas and produced two which present a vast range of information and statistics in a clear and stimulating way. The German paperback publishers Fischer Bücherei and Deutscher Taschenbuch Verlag (dtv) have achieved remarkably high standards of diagram and map in the paperback format. Text book and technical books have been slower to incorporate well-conceived diagrams, but the situation is changing with organizations such as John Wiley, Westermann, Hachette, the Nuffield Foundation, and Penguin Education responding to the challenge at various levels of educational publishing.

The general attitude to diagrams is changing. Diagrams are being used increasingly in newspapers, business supplements and magazines. As text books become more imaginatively designed, so their diagram approach is being considered. Educational magazines, scientific and popular, are being launched which use many diagrams. The problem of diagrams on television is just beginning to be considered, and the design of weather maps has been one of the first steps in the right direction.

There is a real need for the presentation at all levels of statistical and explanatory information in a way which is clear, accurate, and attractive. By looking through the illustrations, it is apparent which do this, and which are not so successful.

This book tries to give the designer a starting point for the solution of a wide variety of problems. It provides a basic, informative text and a comprehensive selection of existing solutions. It should also stimulate thought about the best method for particular problems. A diagram method which conveys one kind of information may be used for a different purpose. Thus a family tree is adapted to getting over the explanations given on forms (page 127); a divided circle is used for tabulating information (page 54).

Diagrams are not things which happen. They need to be as carefully considered as any other aspect of design. My hope is to get designers to produce better solutions and new approaches to the old problem of communicating clearly and with interest.

1 Statistical diagrams

Statistics can be shown visually in various ways. One of the most common methods is that using the graph. The graph contains many of the problems of the statistical diagram and is therefore looked at in detail.

Graphs show how the changes in one quantity are related to changes in another quantity. Common examples are the patient's temperature graph, diagrams showing unemployment figures, movements in share prices. The one quantity (temperature, number of unemployed, price) is related to another quantity (time). The changing quantities are called variables and may be further defined as independent or dependent variables. The independent variable changes regularly and may be a series of equal time divisions. The dependent variable changes irregularly, and represents quantitative or percentage value, such as temperature, number of unemployed or price.

When putting information on a simple line graph, the independent variable is usually calibrated on the horizontal scale (axis) and the dependent variable on the vertical scale (axis). The graph itself is then a series of values plotted along two scales. The name 'line' comes from the result of joining the dots. This line emphasizes the relationship of the variables and can be delineated either by a curved line which passes smoothly through all the plotted points, or by means of a straight line between the points. The line 'catches' the eye and shows the maximum and minimum values, as well as the general direction of rise or fall. The smooth curve should be avoided where possible, since the position of the line often gives a reading which is not based on available information. Continuous readings, such as weather records, made by thermographs, are true line graphs, since information is given for every point of the line.

The use of both vertical scales and marks to label the important points of reference will enable the persistent reader to use a ruler and find out the values.

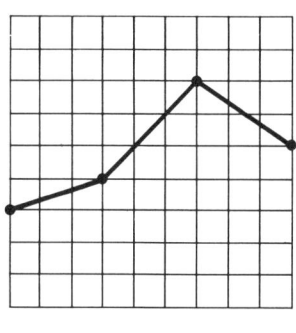
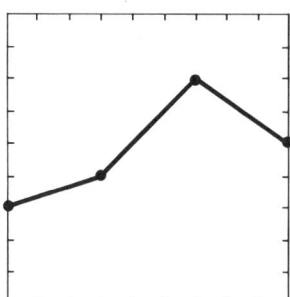
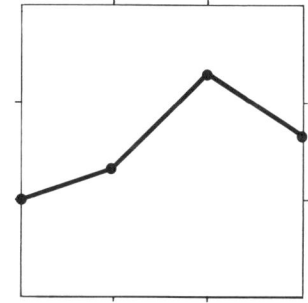
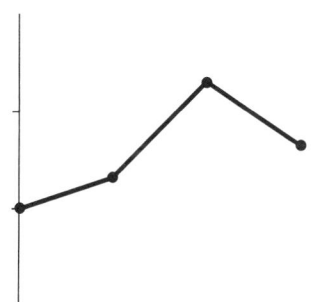

The question of the amount of information a graph should carry is open to discussion and must depend on where it is to be used, what size it is to be, and at what audience it is aimed. The number of curves a graph should ideally contain also depends on its use, size and particularly on the clarity of the coding system; whether broken line and/or colour is used. There is a definite limit to devising broken lines which can both plot values and be read clearly (page 11).

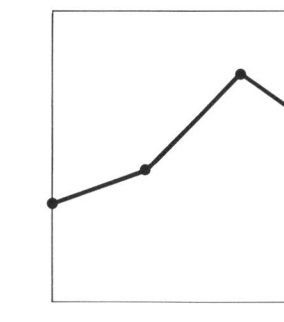
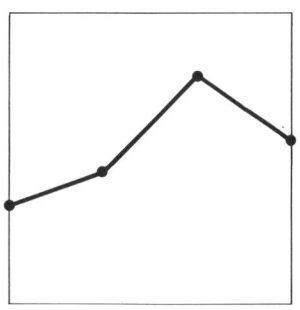

Graphs are usually constructed on graph paper, though this is normally 'forgotten' and the graph published with a blank ground. If the graph paper ground remains, its heavy texture is inclined to detract from the main graph lines. The tendency now is for as few lines as possible. The scale of the dependent variable may have so few indications of values that the reader can only guess at the figures from which the line is plotted.

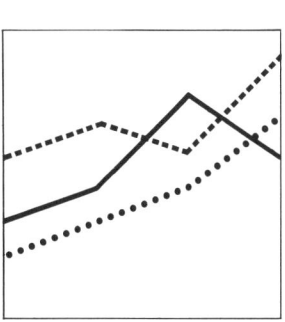
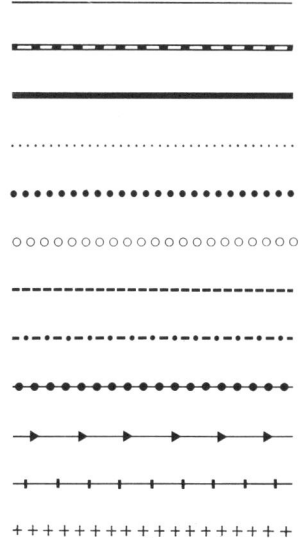

The graph is a widely used diagram, but there are problems which must be avoided. Graphs can be distorted by varying the relationship between the vertical and horizontal scales.

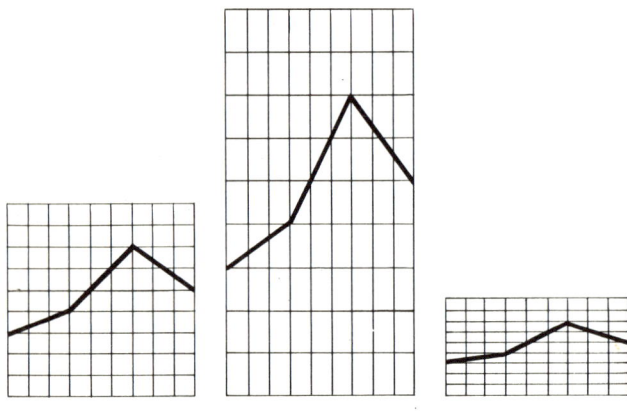

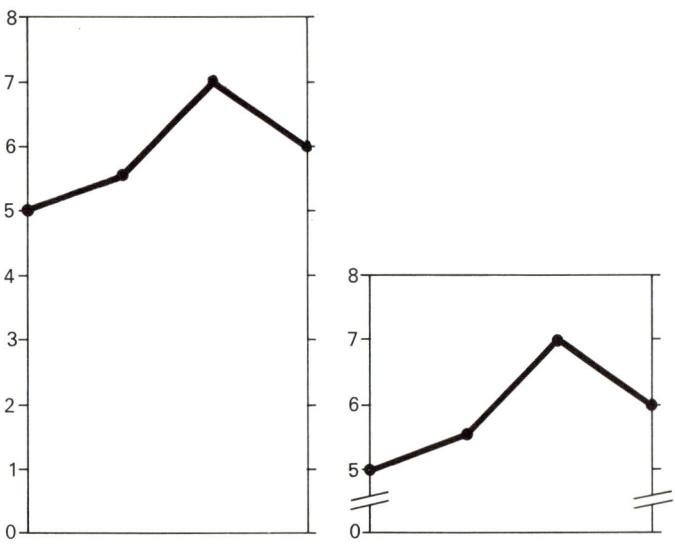

Here the distortion is obvious, but applied in certain circumstances without the advantage of comparative examples, the rise in, say, profit or production could be made to look outstandingly good.

If several curves are included on a graph they must be clearly distinguished so that there is no possible danger of a misreading.

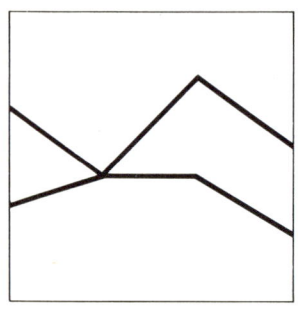 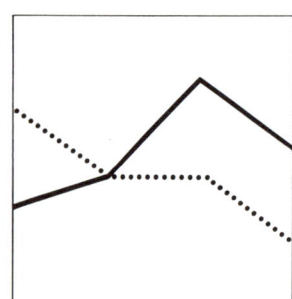

The omission of zero can also result in a distorted impression. If it is a question of getting a diagram to fit a certain depth, this can be overcome by the convention shown above right.

The use of a logarithmic scale produces a different effect and may give the reader a misleading impression. This is discussed on page 20. Two scales may be used on the same graph, one for the left hand and one for the right hand vertical axis. Separate curves are plotted according to their own scale. A direct comparison on the same graph is possible and avoids two graphs side by side. If the lines and scales can each be colour coded this prevents any confusion. An example of this is shown on pages 17 and 38.

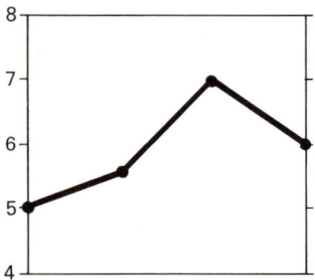

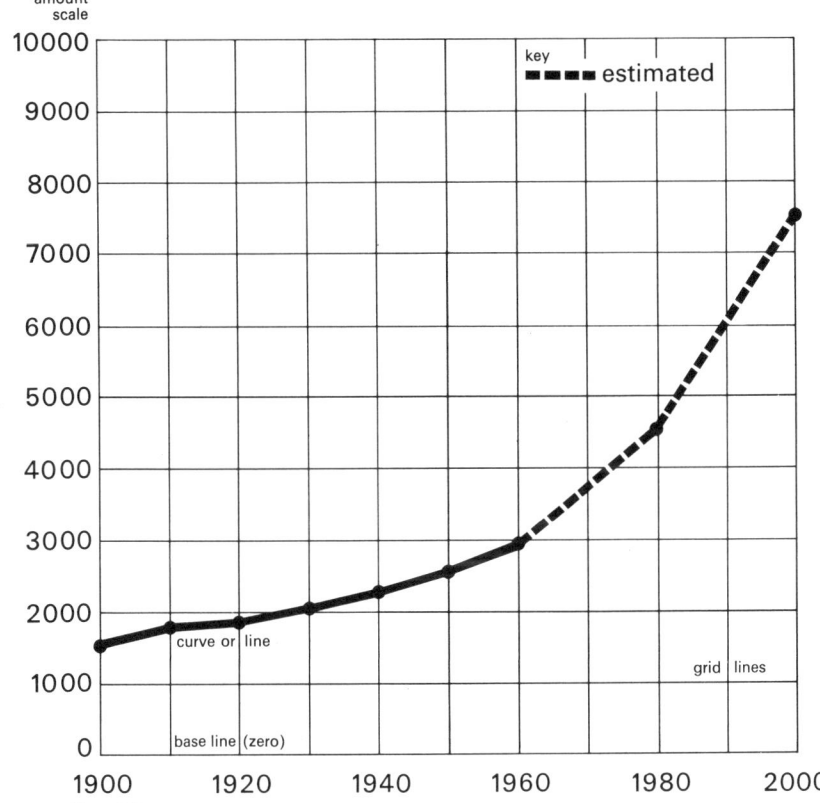

Line graph

Title Ideally the title should be included at the top. It should describe the chart simply and briefly. A subtitle and/or explanatory note may be added. Present practice is to drop the title and put it underneath as part of an explanatory caption. But if the diagram has to stand by itself, a title should be included.

Grid Both the vertical and horizontal axes should be clearly labelled. The amount of labelling will depend on the size at which it is reproduced, the degree of accuracy required, the amount of information available. No more lines should be used than are necessary to guide the eye. On small charts, marks may be used. Lines should be kept light in weight, though the outside line or base could be slightly heavier.

Curve or line This should be the thick line – attracting attention and giving the information quickly. If there is more than one curve, the lines must be distinguished by colour or by various types of dotted and broken line. Dots or some other distinguishing symbol can be used to show clearly the exact point that has been plotted.

Key This is necessary to differentiate the various lines. It should be placed so that it does not interfere with reading the information on the graph. A label clearly related to the relevant curve can be used to avoid a key.

Scale The quantitative scale should start at zero, if changes in values are to be seen in true proportion. If a large part of the grid is omitted the graph can exaggerate small changes so that it is impossible to get any idea of their significance. The scale should be kept simple and the scale caption made to contain the information. Too many zeros should be avoided. A scale reading 0, 1, 2, 3 etc. is easier to read than 0, 1,000,000, 2,000,000, 3,000,000 etc. The omission of zeros must be shown in the scale caption thus: Population (in millions).

Source The source of the data should if possible be stated, especially for an academic audience. This is very seldom included and for the designer presents problems of additional labelling which detracts from the main message of the graph. The solution is a separate list of sources for diagrams.

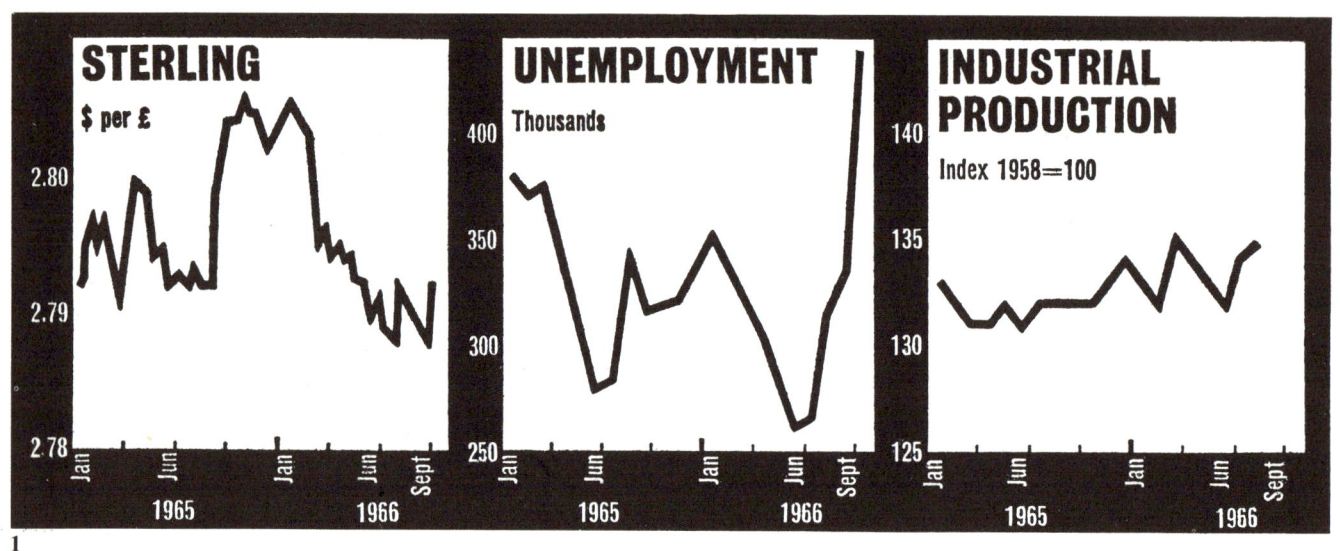

1

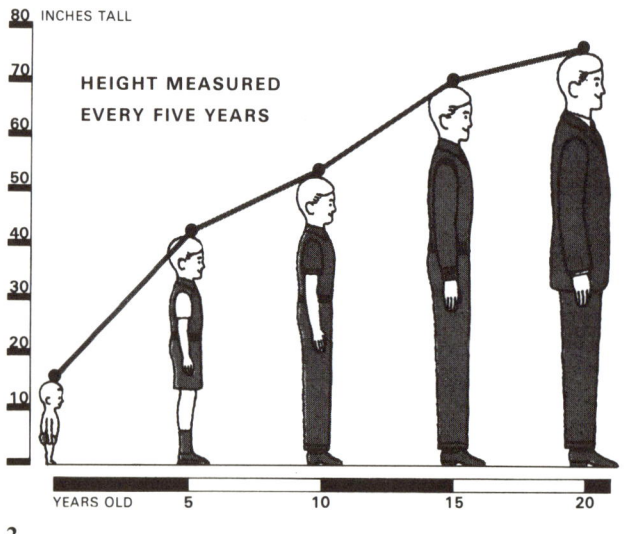

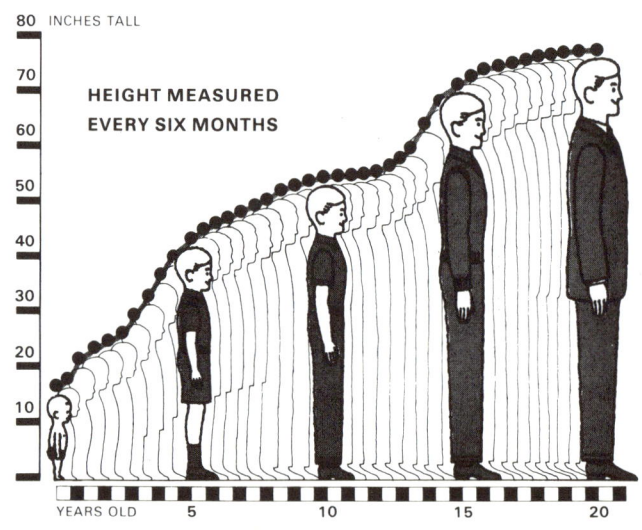

2

3

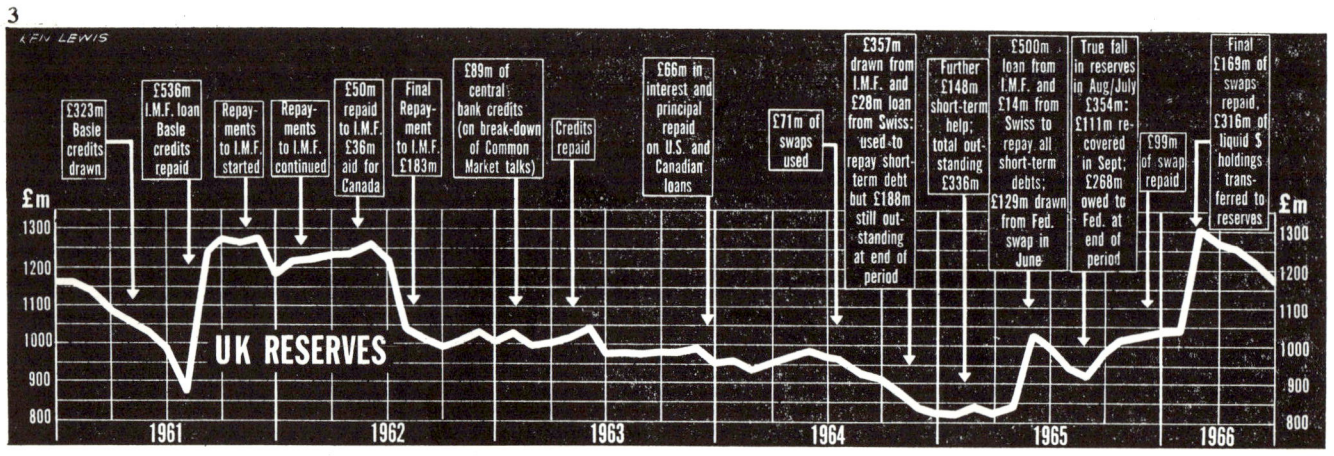

Line graph

1 Three line graphs from a business article, 'The state of the nation', in a newspaper. Impact rather than information has been aimed for, and the labelling is not easy to follow. Yet the graph lines clearly get the message across. *Observer*

2 Two line graphs from an educational book, showing the different graph lines resulting from plotting every five years and every six months. *Mathematics,* Life Science Library

3 Simple line graph from a financial article in a newspaper, showing UK currency reserves. Reduced from 10″. The problem of labelling various key points in the graph line has been neatly solved. *Sunday Times*

4 Graph from a newspaper, showing difference between National Opinion Poll and Gallup result in recording Labour and Conservative support in UK. Lines, clearly coded, 'float' above or below zero (prominently shown!) in centre of graph. *Guardian*

5 One of a series of graphs from a scientific house-magazine, showing fibre stress-strain curves. Each line is labelled and avoids use of a coding system. *Ciba Review*

6 Graph from a paperback encyclopedia, showing percentage of unemployed in relation to number of employable people for various countries. Are six lines on one graph too many? *dtv-Lexikon,* Deutscher Taschenbuch Verlag

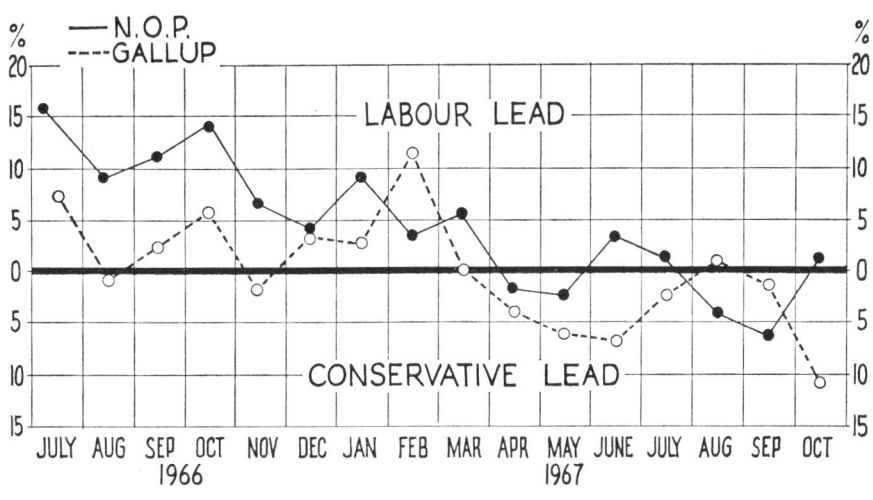
4

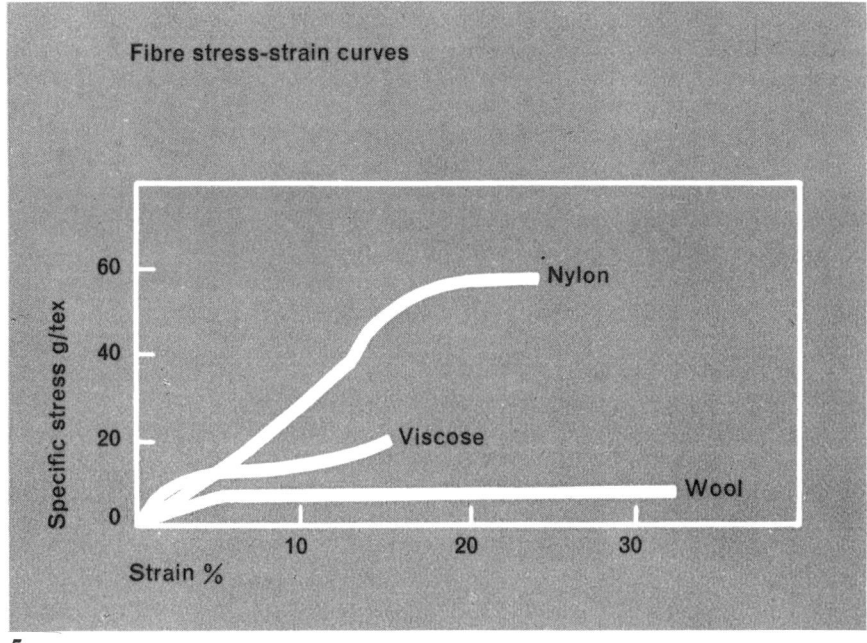
5

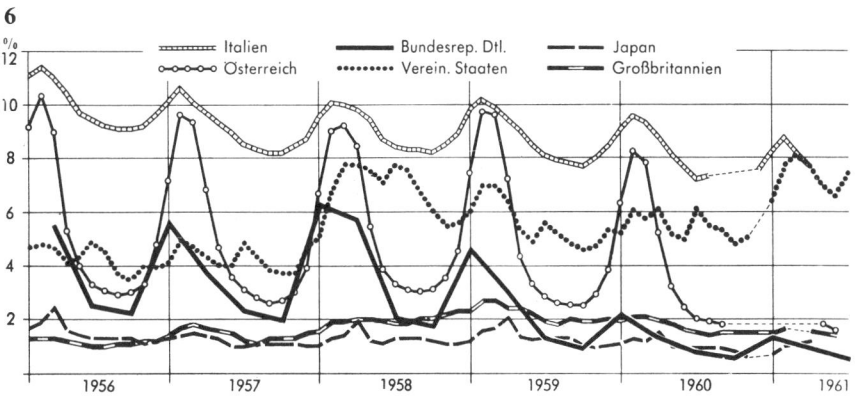
6

Line graph

1 Graph from a scientific magazine, showing history of aviation in terms of speed. It shows the time lag between a record speed and that speed becoming commercially available for transport. Dots suggest the scatter graph (page 22). *Science Journal*

2 Temperature graphs from a geography text book for three locations. Graph records temperature for two months of the year. Shading shows the differences between the three locations.
A. N. Strahler, *Physical Geography*, John Wiley.

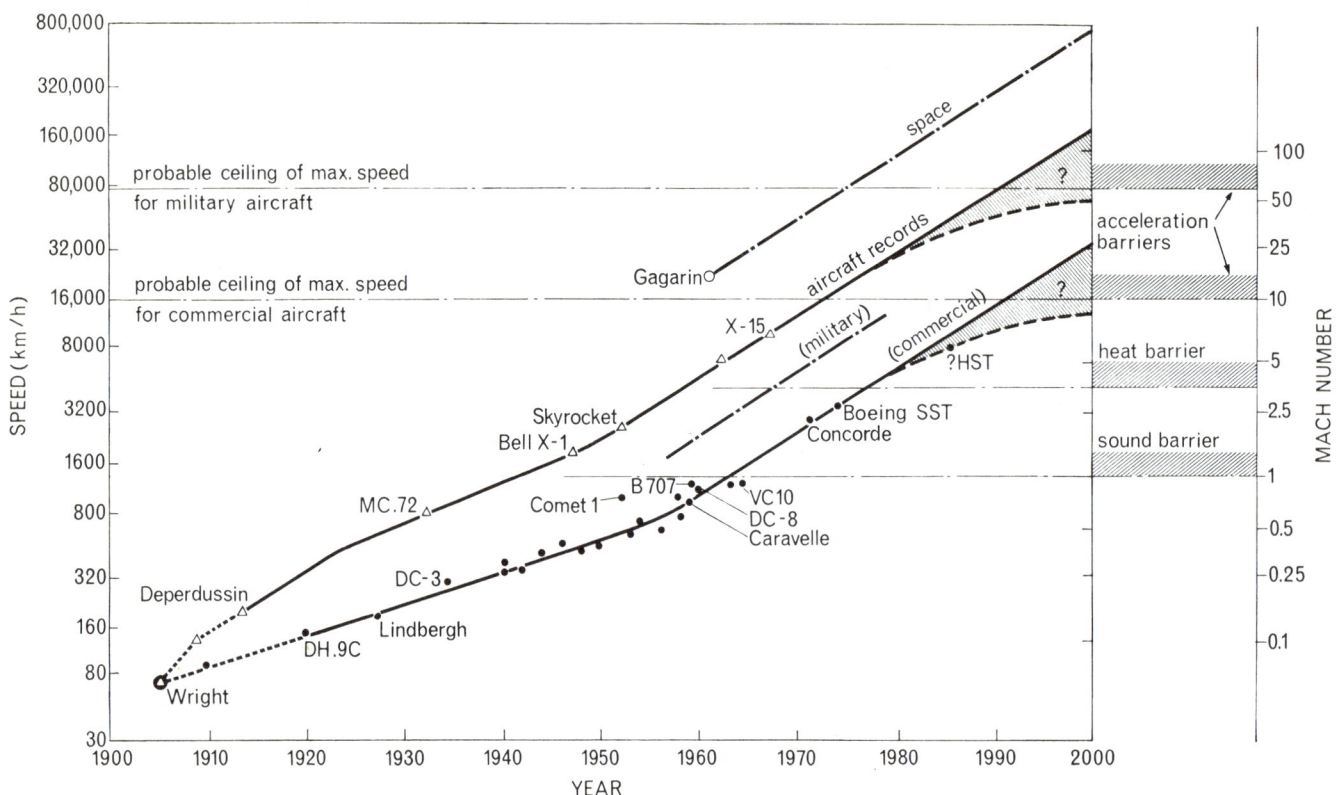

1

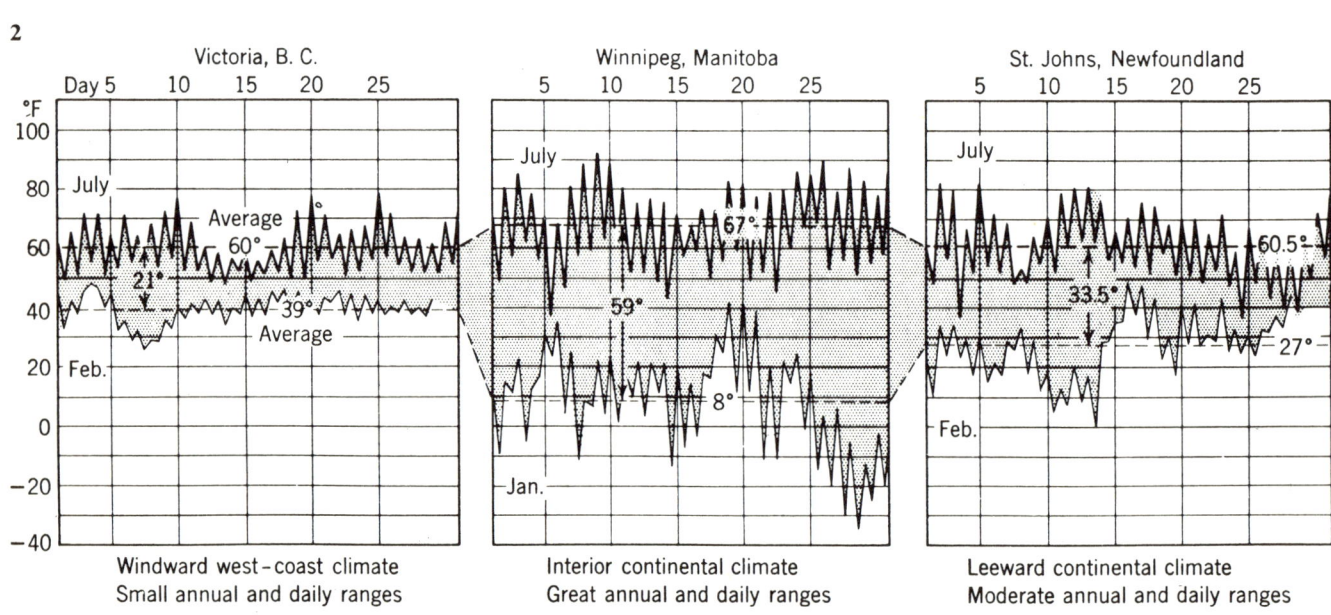

2

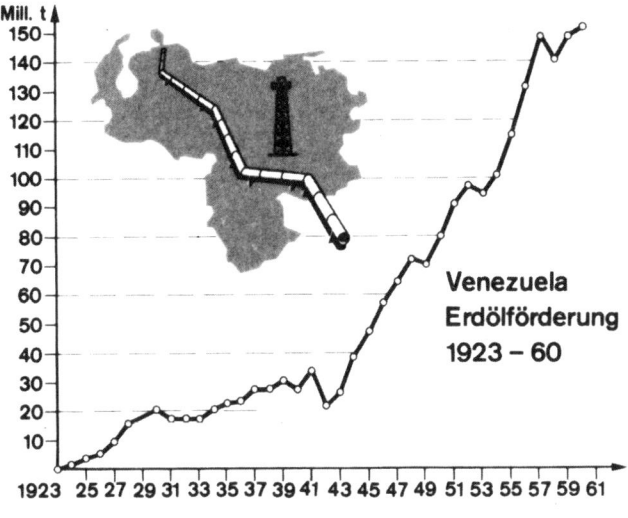

3 Graph from a school text book, showing oil production in Venezuela. Horizontal lines help to read quantity – should there be vertical lines to read the dates? Map, pipeline, and oil rig are an excellent way of immediately conveying the subject of the graph to young readers. A. Jentzsch & J. Winkler, *Länderkunde*, Westermann.

4 Graph from a book, showing town and country population of United States (see page 62). A successful solution to some of the disadvantages of the graph, lack of appeal (overcome by the use of human symbols), 'thinness' (overcome by using different textures to increase weight of lines). This is a divided graph (see page 18) showing the total divided into parts – but it would not be possible to draw it in the conventional way, owing to lines crossing over. Georg Borgstrom, *Der Hungrige Planet*, Bayerischer Landwirtschaftsverlag

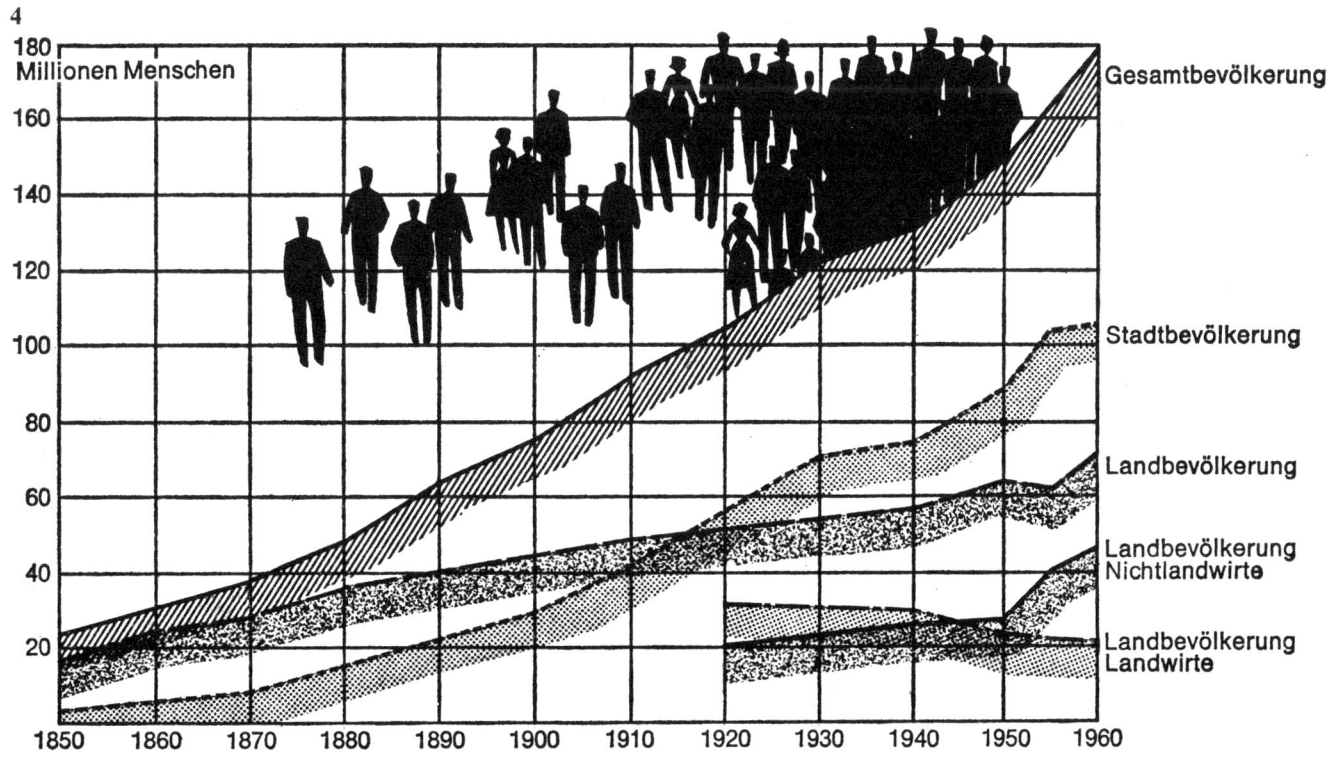

Graph

1 Diagram from a newspaper, showing 'rise and fall of temperature' in the Cold War. The graph normally used to make quantitative comparisons has been adapted to plot events. It requires judgement regarding the event, but this is an excellent solution to a newspaper's problem of presenting events visually as well as verbally. *Observer*

2 Graph from a newspaper, showing play in a rugby match. Position of the ball was plotted every 30 seconds. It shows New Zealand domination of the first half. Again, a non-quantitative statement with location/time information being shown. *Observer*

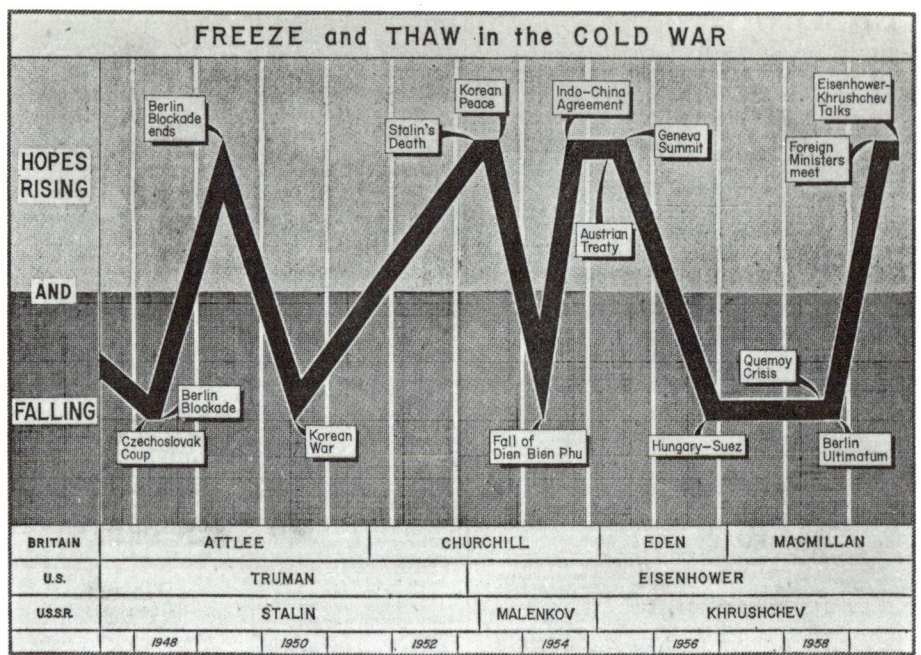

1

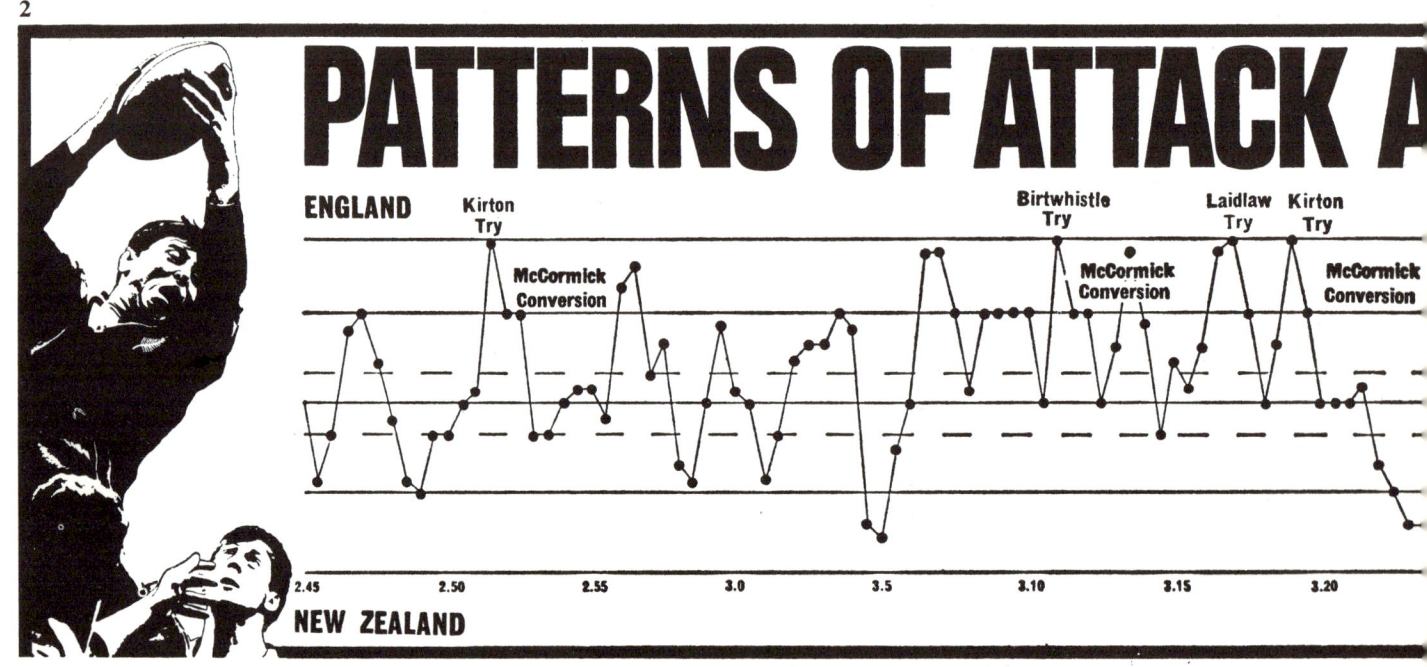

2

3 Graph showing a London Underground timetable. Distance and station locations are shown vertically on the graph, the time scale horizontally. The train paths are drawn in, as diagonal lines between stations (the angle varying according to speed), and as short horizontal lines at stations. This kind of timetable is particularly useful for lines carrying trains of varying speeds and stopping characteristics, especially where there are parallel fast and local tracks which enable non-stopping trains to overtake stopping trains. *How the Underground Works,* London Transport

4 Part of a graph from an educational book, showing generation of electricity, using tidal energy and a single basin or reservoir. Thin graph curve shows high and low tide patterns. Thick line shows height of water in basin. The vertical divisions indicate the operations that are necessary to generate electricity and when they take place. E. G. Sterland, *Energy into Power,* Aldus Books

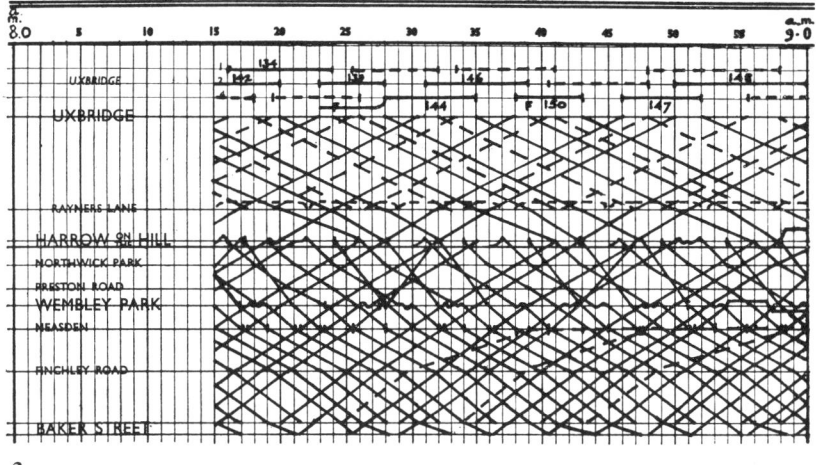

3

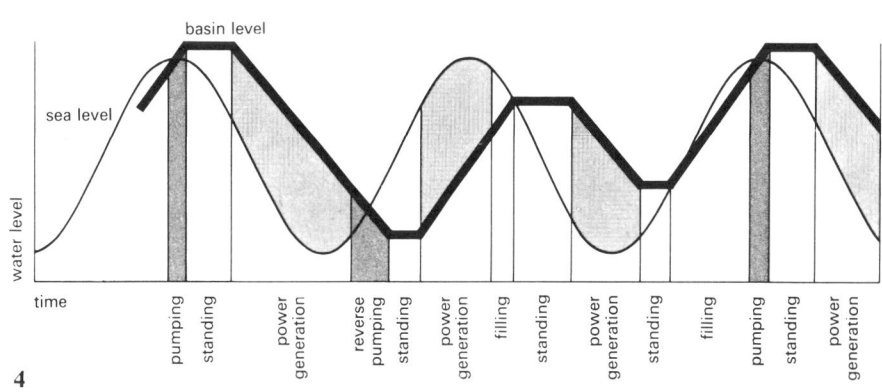

4

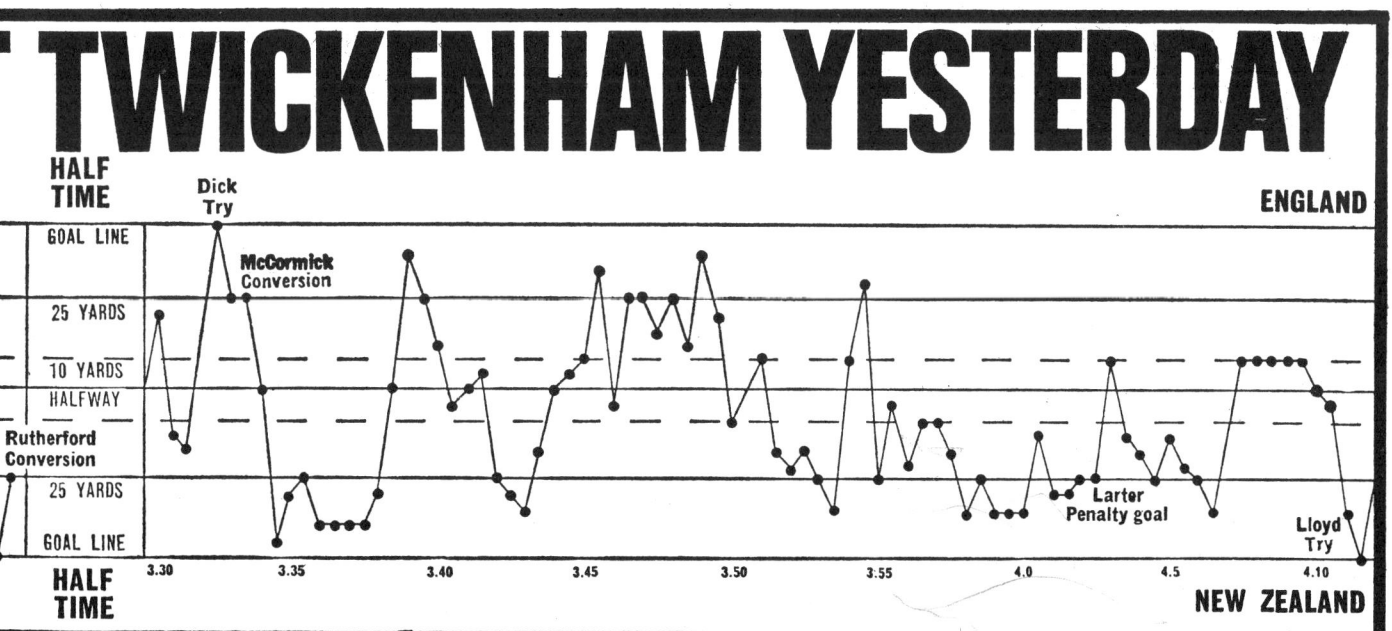

Graph

1   Diagram from a scientific magazine, showing density profile of New York. The three curves are for 1920, now, and 1985. They illustrate the changing density level and location of the population. The graph has been used as a more suitable alternative to plotting the information on a map. *Scientific American*

2   Graph from a text book showing wine production in France. This shows the use of *moving averages* which smoothes out a graph line that seems to have no direction, caused by widely varying individual values. This is a specialized technique which aims to show a general trend difficult to see from normal plotting methods. F. J. Monkhouse & H. R. Wilkinson, *Maps and Diagrams,* Methuen

3   Graph from a scientific paperback showing the population curve (black) related to total water consumption curve (red) for New York. The use of two scales. Letters give explanation for variations in curve, e.g. C: Population shift into armed forces and outside industries during World War II. Michael Overman, *Water,* Aldus Books

4   Graphs from an atlas showing world production of pig iron/crude steel. Crude steel is shown as a surface graph (area shaded by tone) and pig iron as a line graph. The two elements sorted out by use of line and tone. *Shorter Oxford Economic Atlas,* Oxford University Press

5   Graph from a scientific magazine showing US annual energy consumption of various energy sources: wood, hydro-electricity, gas, oil, and coal. Use of third dimension to differentiate and dramatize information. *Power*

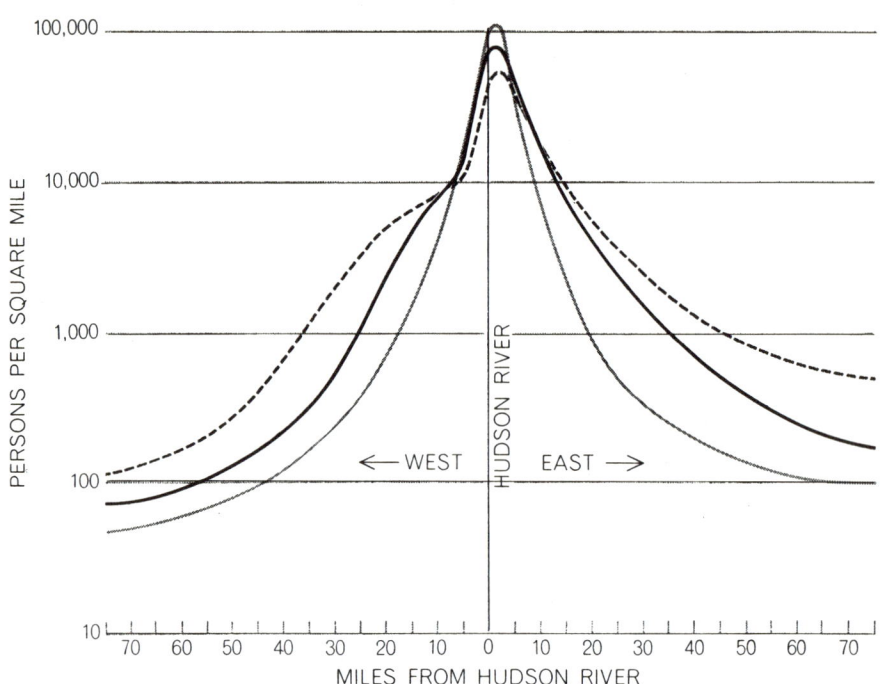

1

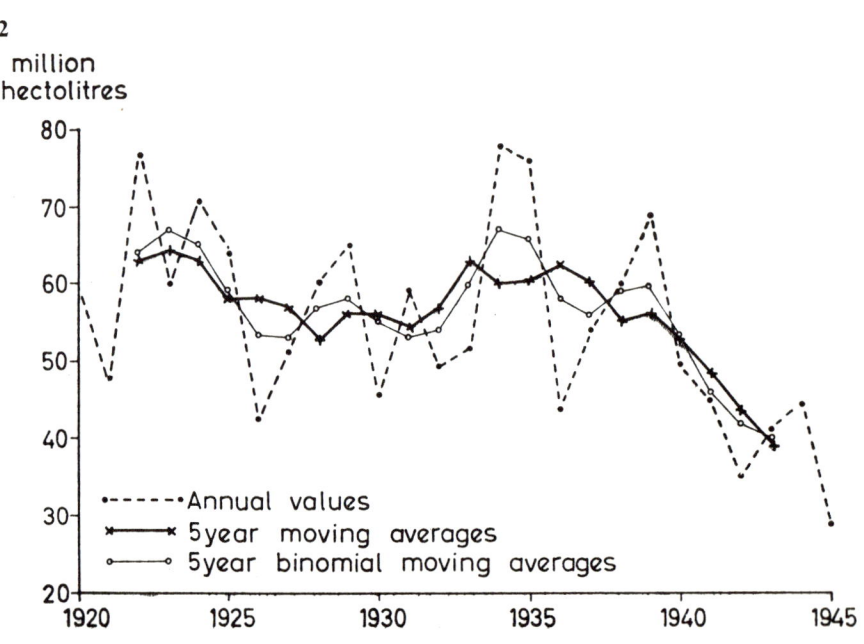

2

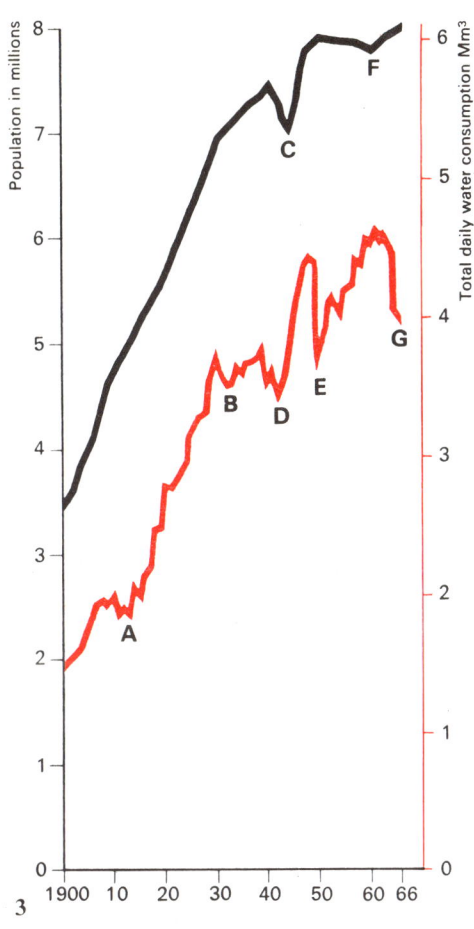

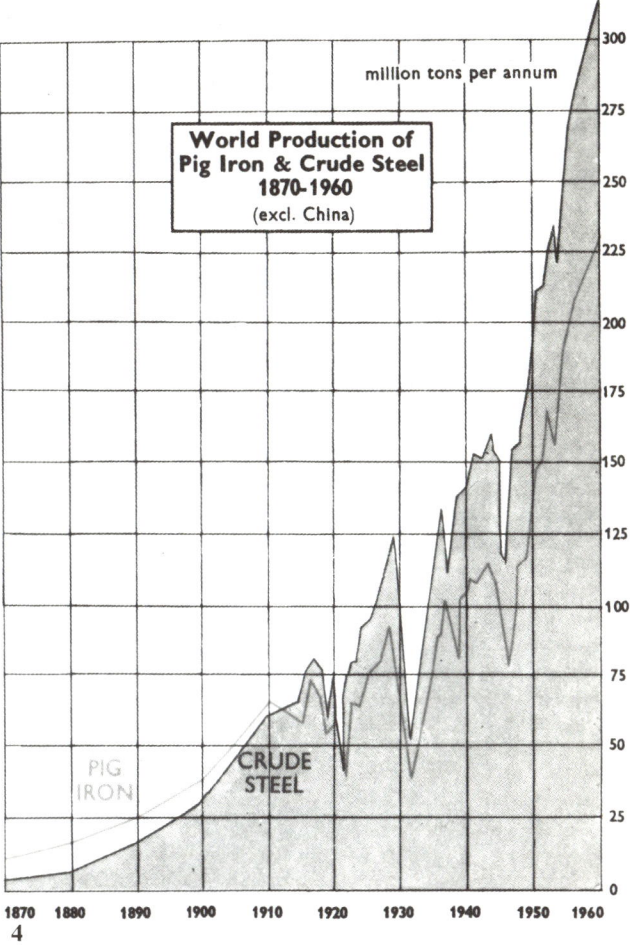

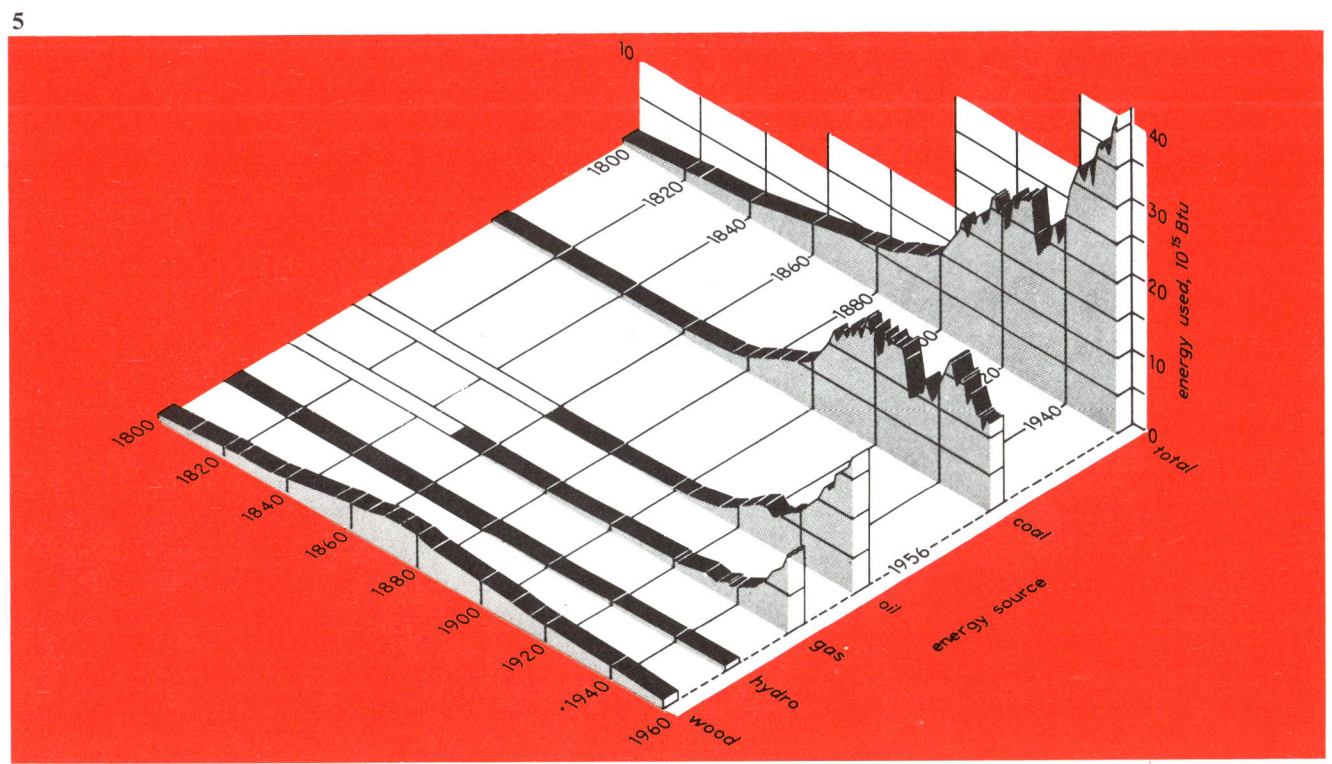

Divided (compound) line graph

The divided or compound line graph shows the value of both the total and its parts by a series of lines on the same frame. Provided the vertical scale begins at zero, it is possible to divide the area beneath the main line into any number of parts. The area between the successive lines should be shaded distinctively.

Sub-division should be attempted in moderation. It is not easy on divided graphs to follow increase or decrease in all the parts. Ideally the more stable elements should be at the bottom to show clearly the variation in the other elements. But this may not always be possible.

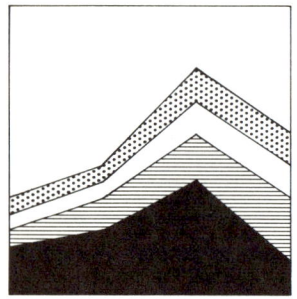 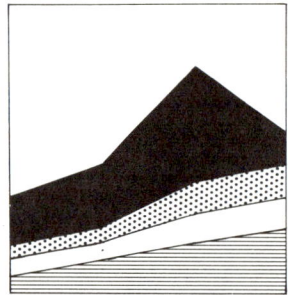

The problems for the designer are using a satisfactory system of textures and labelling the various parts clearly.

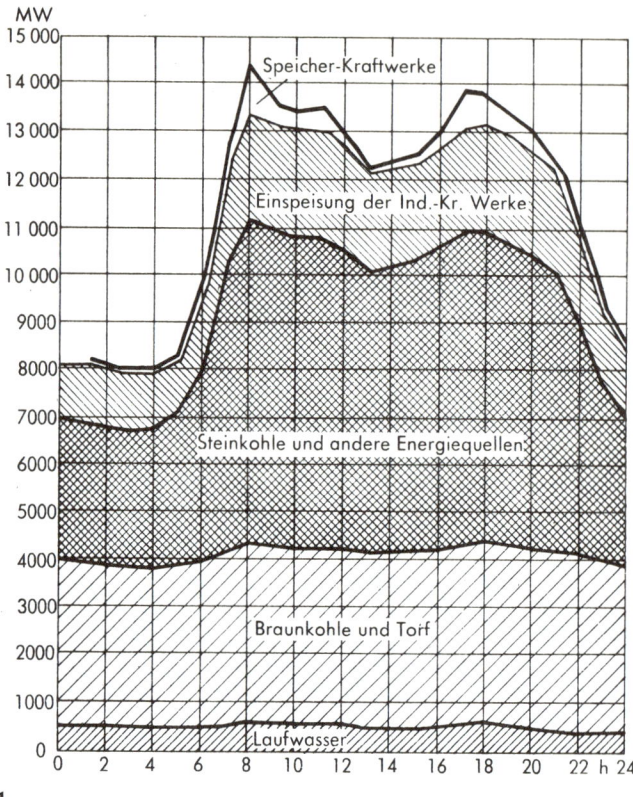

1

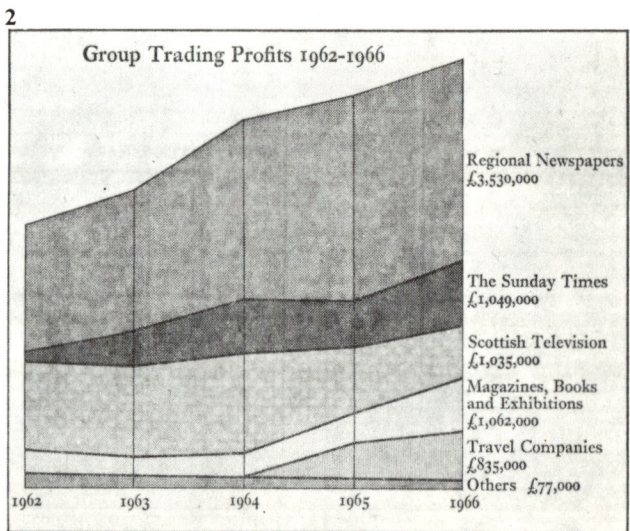

2

18

1 Graph from a scientific paperback, showing the use of different kinds of power station during one day. Labelling on the area has problems owing to texture or size. *Das Fischer Lexikon, Technik*, Fischer Bucherei

2 Graph from a newspaper advertisement showing trading profit of a publishing group. Lettering at the side solves the labelling problem. Reduced from 5½″. *Thompson Organisation*

3 Graph from a scientific magazine showing change in occupations of the labour force. The general trend of more people providing services and fewer people producing goods is clearly shown but the change in individual parts is not always clear. Reduced from 7″. *Scientific American*

4 Graph from a business magazine, showing rise in world population. A dramatic and original solution (in full colour occupying complete page) ideal for a magazine but not for the statistician. Compare with other solutions of the same problem, page 55 and 57. *Fortune*

3
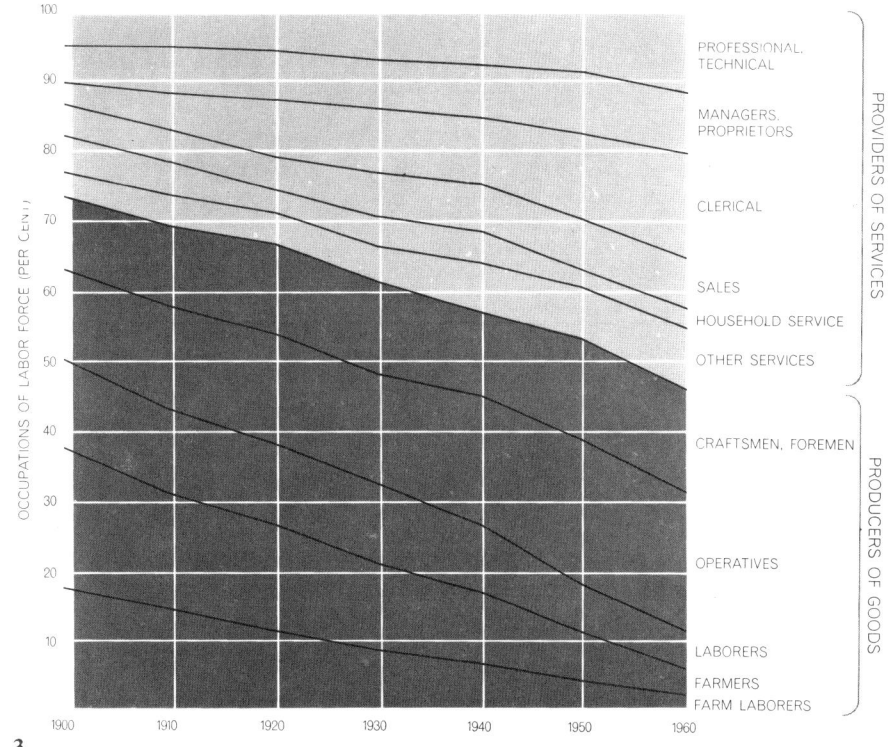

4
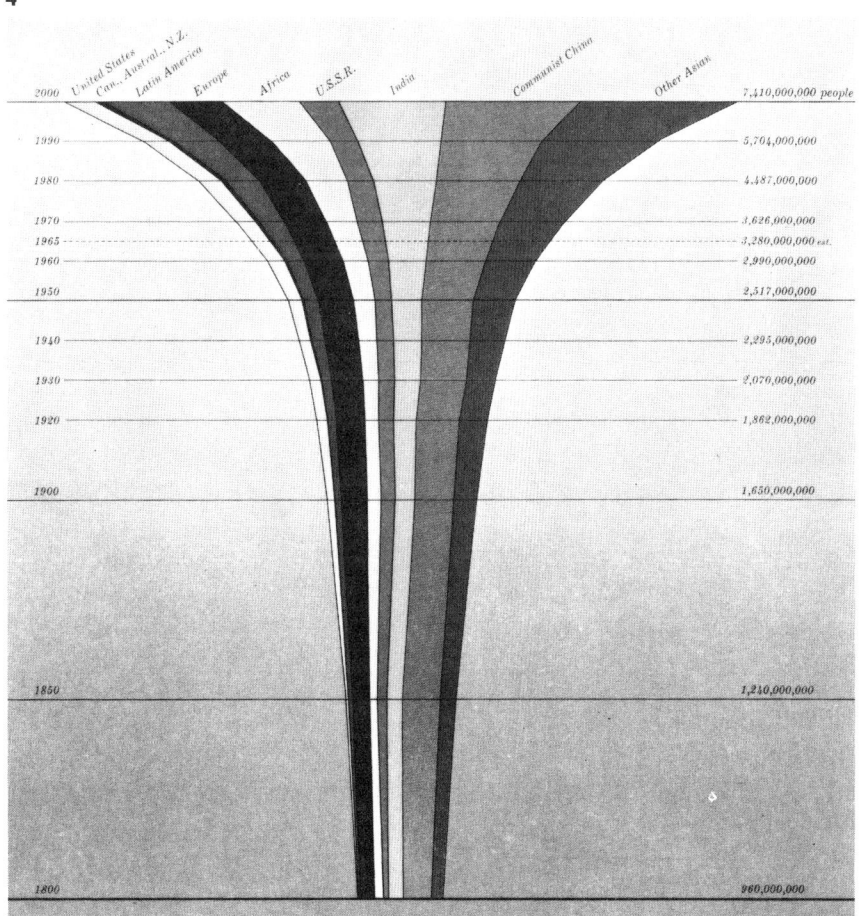

Logarithmic graph

The line graph shows changes in quantity. It plots exact increases or decreases. But this is not always the information which is needed. Sometimes the rate of increase rather than the amount is wanted. The rate of increase as well as the amount can be shown by using a logarithmic graph.

The scale on a logarithmic graph appears distorted; the vertical scale is not evenly spaced. The numbers on the scale are spaced equally but refer to the logarithmic factors, thus not providing a linear progression. This results in open and grouped lines on the vertical scale. Each of these groupings is called a cycle and the top and bottom of each cycle (where the logarithmic factor is 10) must be 10 or a decimal or multiple of 10, e.g. .01, 100, 1000.

To see how the logarithmic graph works, the populations of three towns have been plotted on graphs. Left on an arithmetic grid showing absolute gains. Right on a logarithmic grid showing rate of change.

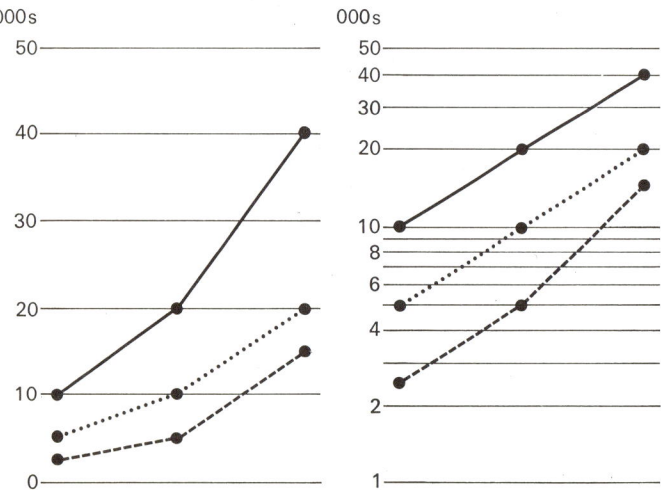

Specially ruled graph paper can be bought, and it is not necessary to work out one's own vertical scale.

The logarithmic graph is especially useful when comparing production figures and the rate of increase. This is important where items vary considerably in the amount produced.

The logarithmic graph is also used when figures need a large vertical range that would be difficult to fit on a normal graph. But it is a specialized device for a particular job. The odd looking vertical scale makes the ordinary reader wary of the information given.

1 Graph from a technical publication, showing world production of major apparel fibres. The use of the logarithmic scale enables silk (very small production) to be compared with cotton (very large production). It is easy to see the rate of increase of rayon. F.A.O. Commodity series, *Fibers*

2 Graph from a consumer magazine showing how share values have risen. The logarithmic scale enables the depth of diagram to be restricted. 'Money' *Which?*

3 Graph from a scientific magazine showing energy consumption in past and suggesting future trend. Again, easy comparison of small and large figures is possible, as is also the probable rate of growth. Unfortunately the equally spaced horizontal grid lines do not immediately indicate the kind of scale used. *Power*

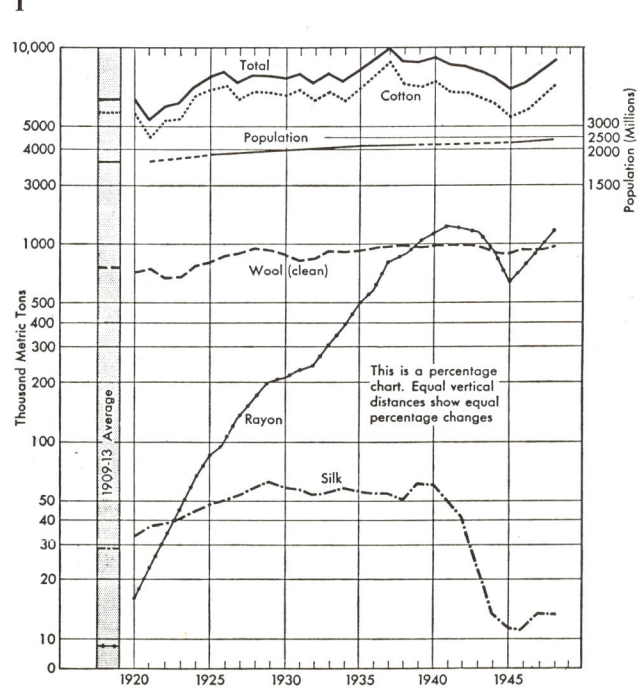

1

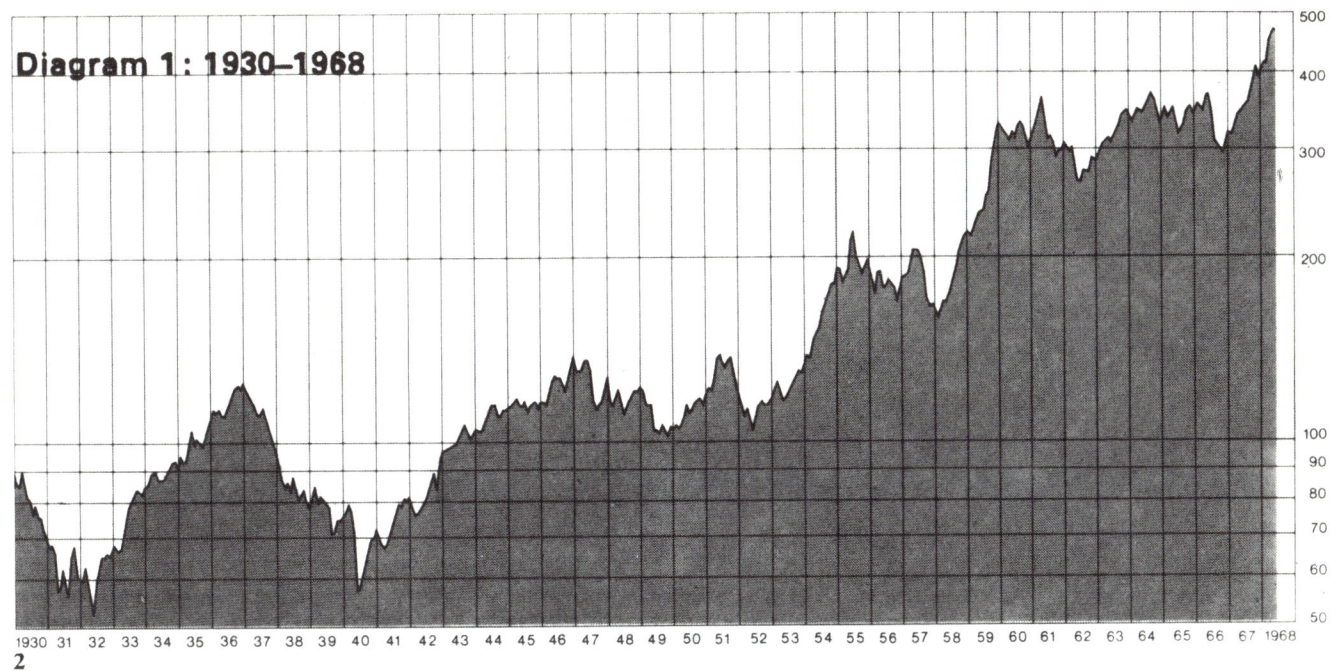

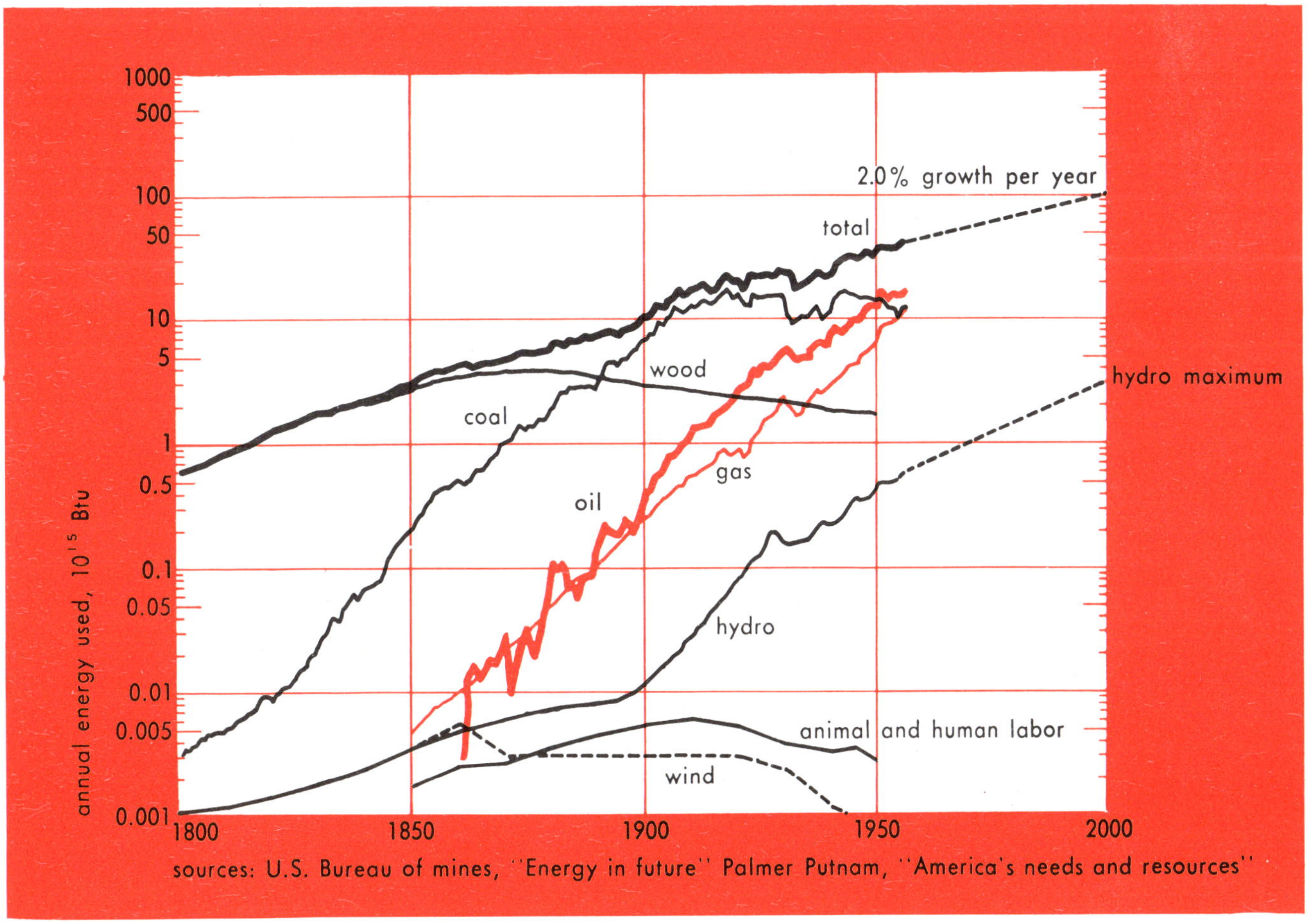

Scatter graph

The scatter graph, like the logarithmic graph, is a specialist tool, and as such is not used widely. The basic principle is that a single dot to represent one value is plotted against scales drawn on two axes at right angles in the normal way. There is no question of joining the points by a line, each example being separate. The graph consists of a 'scatter' of dots and from this the basic message is read.

The message has to be worked out from the pattern which the scatter of dots shows. If the dots show a random scatter, there is no relationship between the variables. The 'scatter' graph becomes more involved by introducing symbols that differentiate aspects of the information.

1 Graph from a scientific magazine, showing rate of metabolism, providing a limited means of adaptation to cold. The grey tone indicates the general trend of the 'scatter'. *Scientific American*

2 Graph from an educational book, plotting capabilities of plasma containment devices. There is no general trend (aim is indicated by arrow) and dots remain a 'scatter'. The diagram emphasizes that there is still a long way to go before controlled thermo-nuclear fusion is a reality. E. G. Sterland, *Energy into Power*, Aldus Books

3 Graph from a magazine, showing athletes plotted according to shape. William Sheldon devised a system of identifying physique by shape alone, not size. He arrived at three groups of body shape – rotund, herculean, and lean. Since few people are obviously in one group or another, he developed a more comprehensive and complicated classification system to cope. When these results are plotted on a triangular graph, with the extremes at the three corners, a general picture from the dots soon emerges. College students produce a random picture. Olympic athletes produce a compact picture. *Observer*

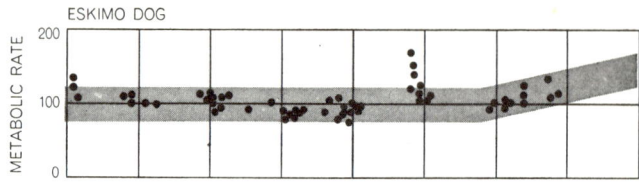
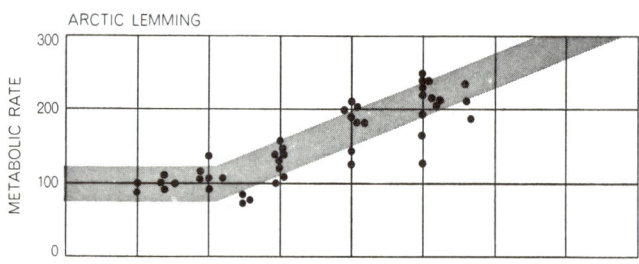
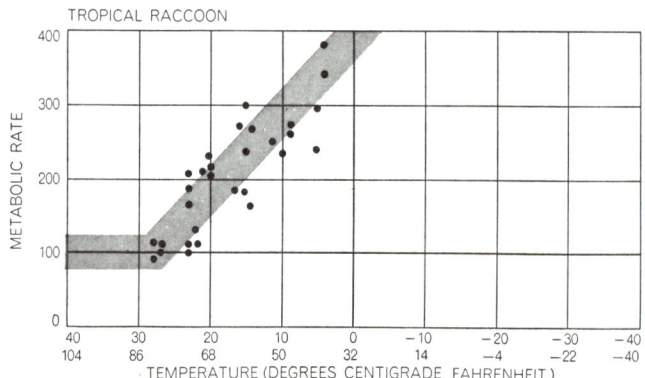

1

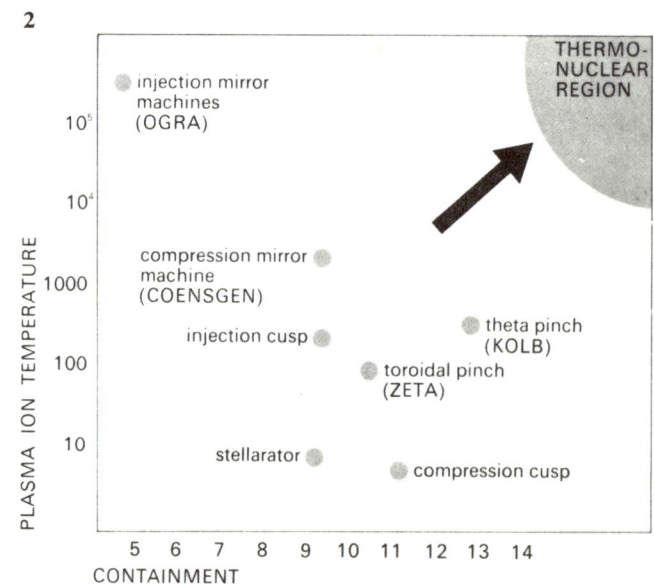

2

To be an athlete does not mean being any one standard shape. Particular physiques tend, on average, to be suitable for particular events. The traditional image of Herculean man is of a man equipped for strenuous labour, yet many athletes who seem hardly 'athletic' at all achieve prodigious feats of endurance. There are three extremes of body shape, and each of us possesses a bit of each. How near we are to each of these extremes defines our shape, and also our physical suitability for a particular endeavour; these can be plotted diagramatically (left and below). There are always exceptions, and many of today's supreme athletes are winning events for which they do not possess the traditional shape and stature. Swimmers can be almost any shape – and win.

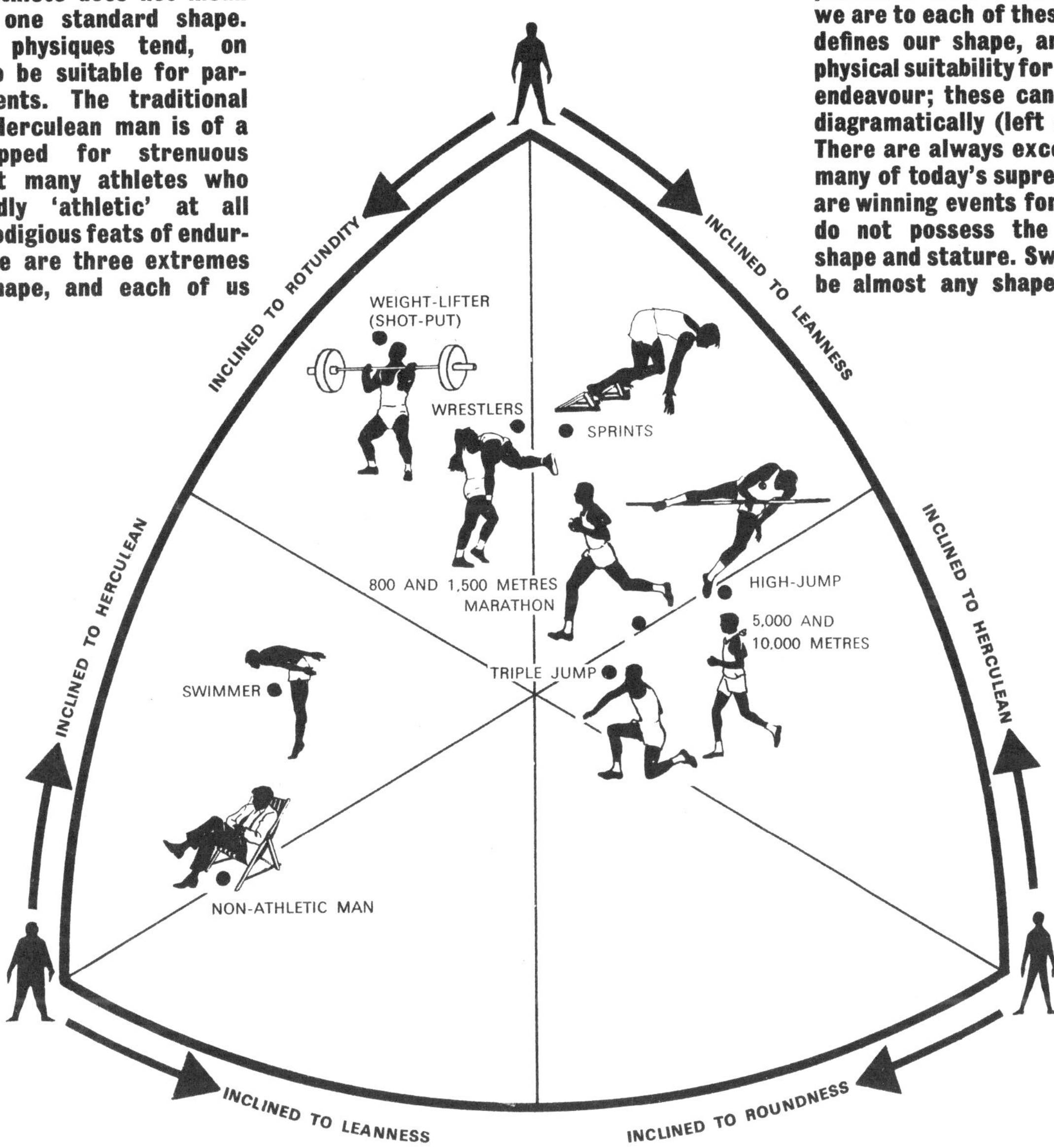

shapes, produce just such a random picture (see diagram right).

The Olympic athletes produce a compact picture. Of the 137 men measured in 1960, all of them were in one half of the graph (see diagram on right). No one who was at the Rome Olympics scored more than four out of seven for endomorphy. The men were either ectomorphs or mesomorphs, or combinations predominantly of these shapes.

In other words, the plump and the round should forget the whole idea of ever being Olympic stars. Schoolboys, told hoarsely that it is just "guts" or "drive" that will win them the race, should examine their shapes seriously, and then yield if they find themselves blatantly on the wrong side of the somatotype graph for that particular event. Their shape is against them.

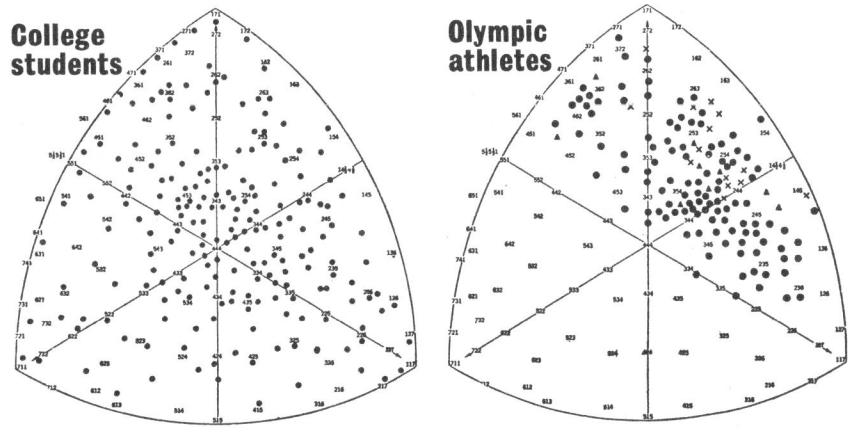

Diagrams based on "The Physique of the Olympic Athlete" by Dr J. M. Tanner (George Allen & Unwin)

Scatter graph

1 Graph from a paperback, showing distances of selected countries from USSR and China. Maps do not always show clearly the sphere of influence of countries, and this kind of diagram is an excellent way of comparing location of areas within the influence of two major powers. For convenience, distances are measured between capitals, which can be misleading. Afghanistan, for example, actually touches both countries, yet Kabul is far from both Moscow and Peking. J. P. Cole, *Geography and World Affairs,* Penguin Books

2 Graphs from a consumer magazine, comparing scores using two different kinds of hearing aids. Position of the dots on either side of the diagonal line indicate the quality of the aid. *Which?*

3 Graph from a scientific magazine showing energy use v. gross national production for 1961. The positions of symbols indicate clearly the industrialized and under-developed countries of the world. *Scientific American*

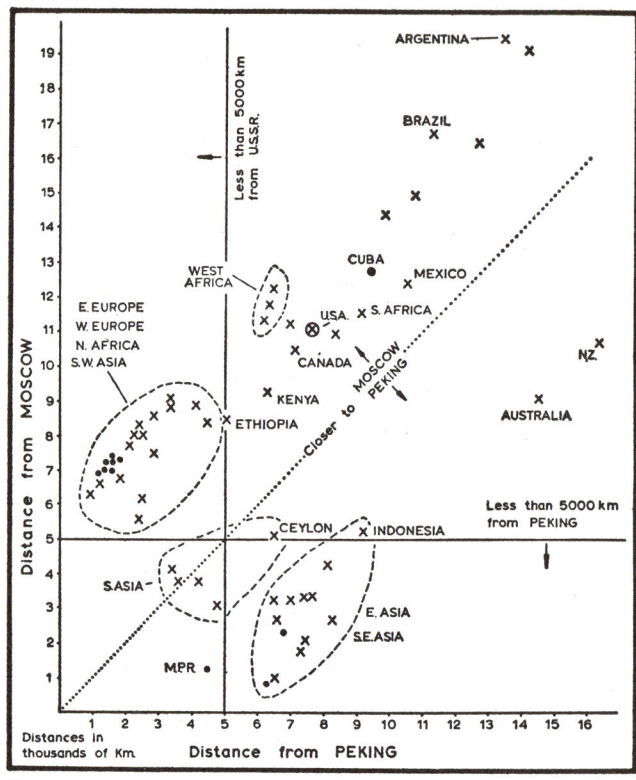

1

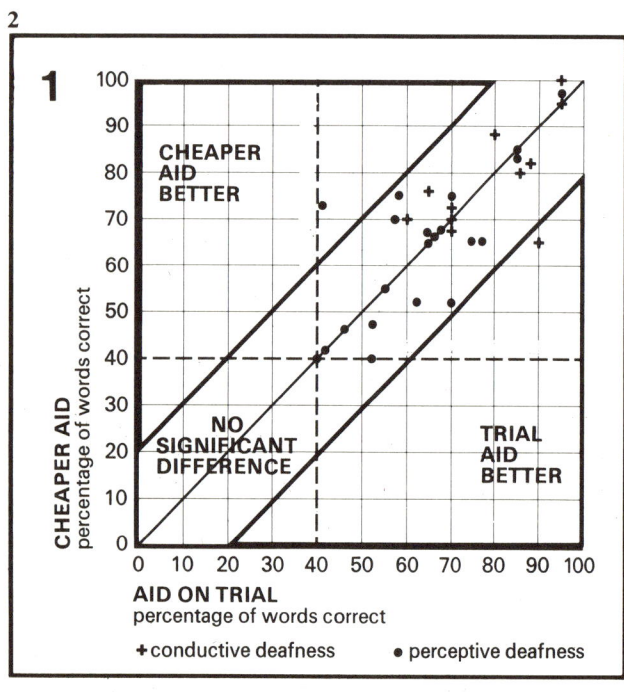

2

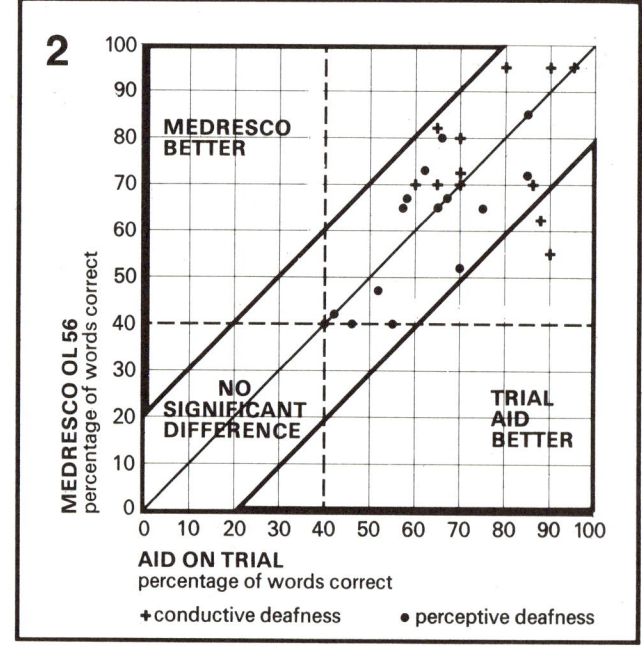

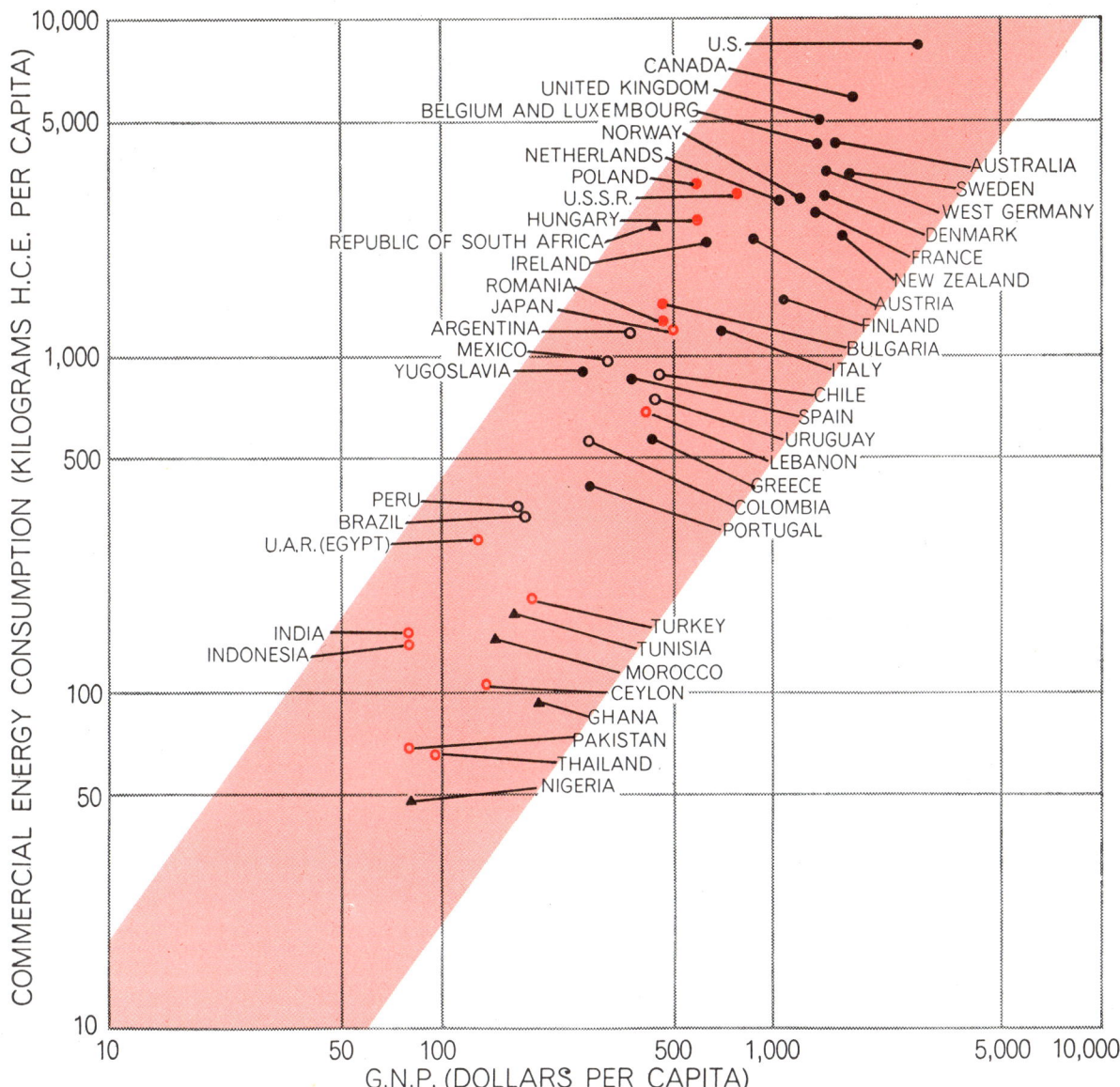

Bar (column) graph

The bar or column graph is directly related to the line graph. It is drawn from a series of values plotted against two axes, but instead of being joined by a line, the values are represented by vertical bars. Each bar is usually kept separate from its neighbour. The bar graph emphasizes the actual quantity (often comparing separate items), while the single line graph shows more clearly the rise and fall of the values of one subject. The visual difference is not hard to see.

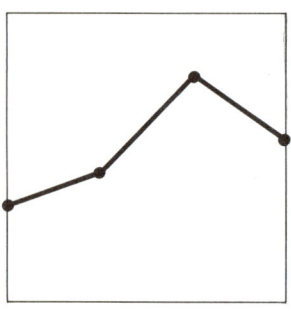
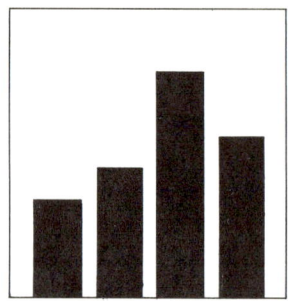

The bar graph has other advantages. It is particularly useful where the information consists of distinct units (months, years). The separate nature of the bars emphasizes the unit. Another important advantage is that bar graphs can be turned on their sides. The main advantage of this is in the placing of lettering for easier reading.

It is possible to show two or more items on the same bar graph either by placing them side by side or overlapping them slightly. Colour, tone, or texture help to differentiate bars. But the more items included, the more difficult it is to follow through the information. The line graph has the advantage in this situation (pages 32–3). But in many cases the bar and line graph are interchangeable.

As in the line graph (and in fact any diagram), labelling can be done in a variety of ways, including too little labelling, which makes information hard to read clearly.

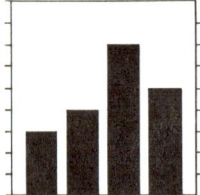
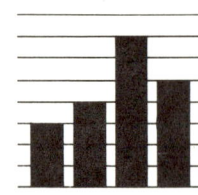
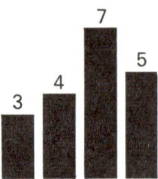

Bars should be of the same width, or the value of the information shown in the same way is lost. There is an exception where, for example, there may be a change in the time scale, i.e. from 5 years to one year. This should be shown by varying the width of the bar accordingly.

Similar to the bar graph is the histogram (see page 38). The term was originally used for diagrams showing amounts of rainfall. It is now applied to diagrams showing frequency distributions and other values. The bar graph shows absolute quantities, the histogram usually percentages. The bar graph consists of separate bars, the histogram usually has no spacing between columns since there is no definite break in the series.

Another variation of the bar graph is the bar line. In this, the coordinate points are joined to one of the axes by perpendicular lines, instead of being connected to each other by a continuous line as in the line graph (see page 39). The line is a substitute for the curve or bar. This may be a useful solution where there are many figures and each needs to be read.

A pyramid form of the bar graph is used to give comparative information about population (see page 36).

1 Graph from newspaper, showing US troops in Vietnam. The problem of coping with very small and very large numbers. *Sunday Times*

2 Graph from an educational book, showing 50 years of pig iron production. Bar graphs can be placed horizontally. Marie Neurath, *Living with One Another*, Parrish

3 Graph from a newspaper, showing UK Christmas road deaths. The alternating black and white has obvious graphic meaning but does not make it easy to read. Reduced from 12″. *Sunday Times*

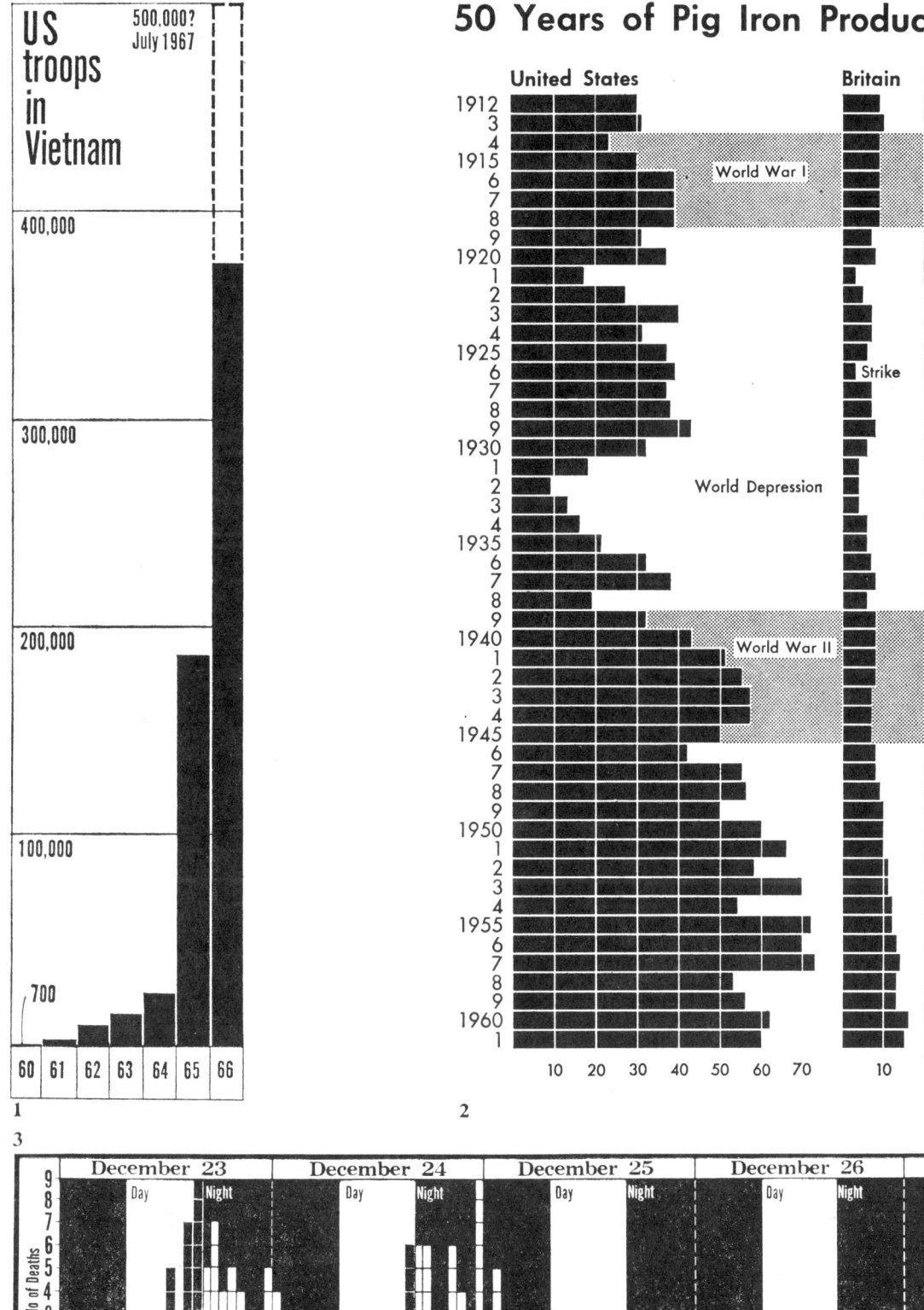

Divided (compound) bar graph

Each bar in a bar graph may be divided into any number of parts to compare constituents as well as total value. The result is known as a divided or compound bar graph. The main problems of this kind of diagram are the number of divisions needed for clear information, the kind of texture, tone or colour to be used to differentiate the parts, and the labelling or keying of the parts.

It is often difficult to follow increases or decreases in the various parts. As with the divided line graph, it is better to try to get the more stable element at the bottom to help comparison of the other parts. Lines can be drawn across from each bar to make easier comparison (see page 46).

1 Graph from a book showing total population of two groups of animals (cud-chewing and non cud-chewing, including man). Symbols identify and 'humanize' the parts. Georg Borgstrom, *Der Hungrige Planet,* Bayerischer Landwirtschaftsverlag

2 Graph from a scientific magazine, showing new thermal-generating capacity 1958–67, broken down into fossil units large and small (coal, oil or gas) and nuclear units. Tone, and particularly colour, code the elements. *Scientific American*

3 Graph from an informational booklet, showing producers of vegetable oils. Textures and a key identify the various parts. US Department of Agriculture.

4 Graph from an educational book, showing hydro-electric power potential (wavy line), and the production that has been developed (black). Simple division into two parts and the impact seems more important than the actual figures. Compare with page 42. E. G. Sterland, *Energy into Power,* Aldus Books

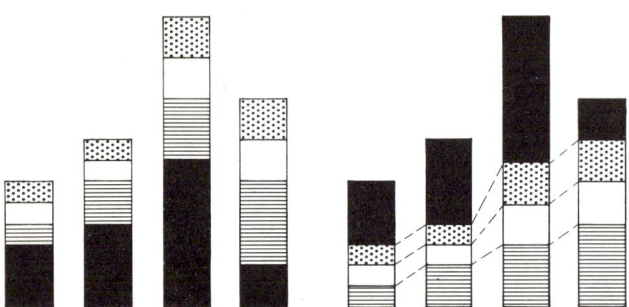

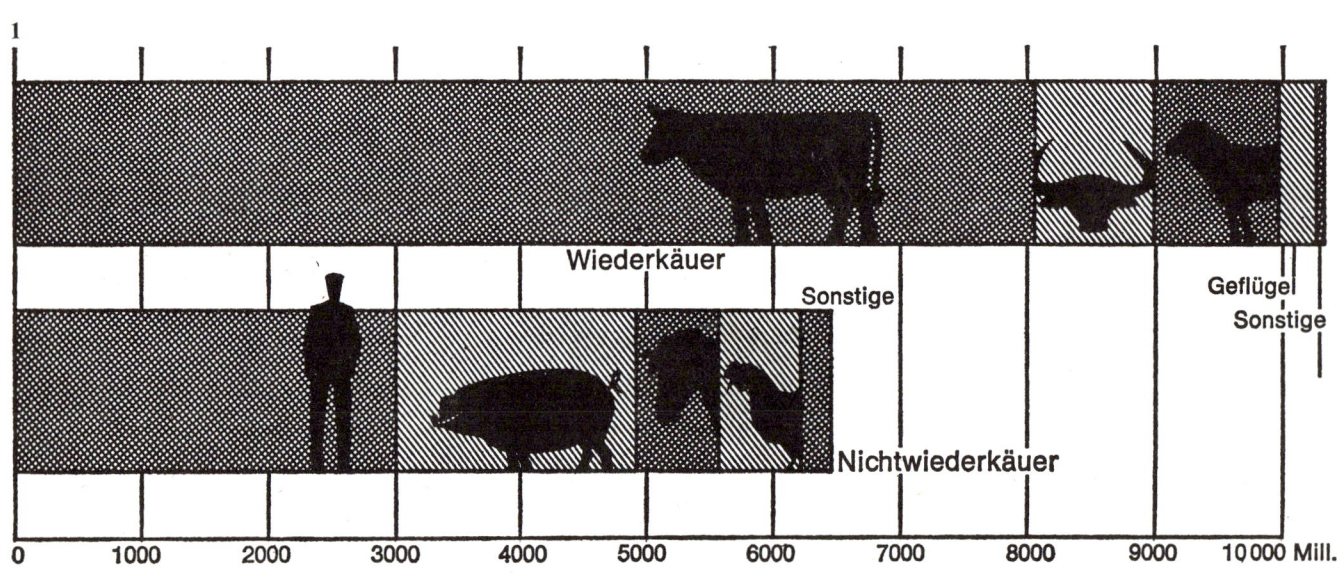

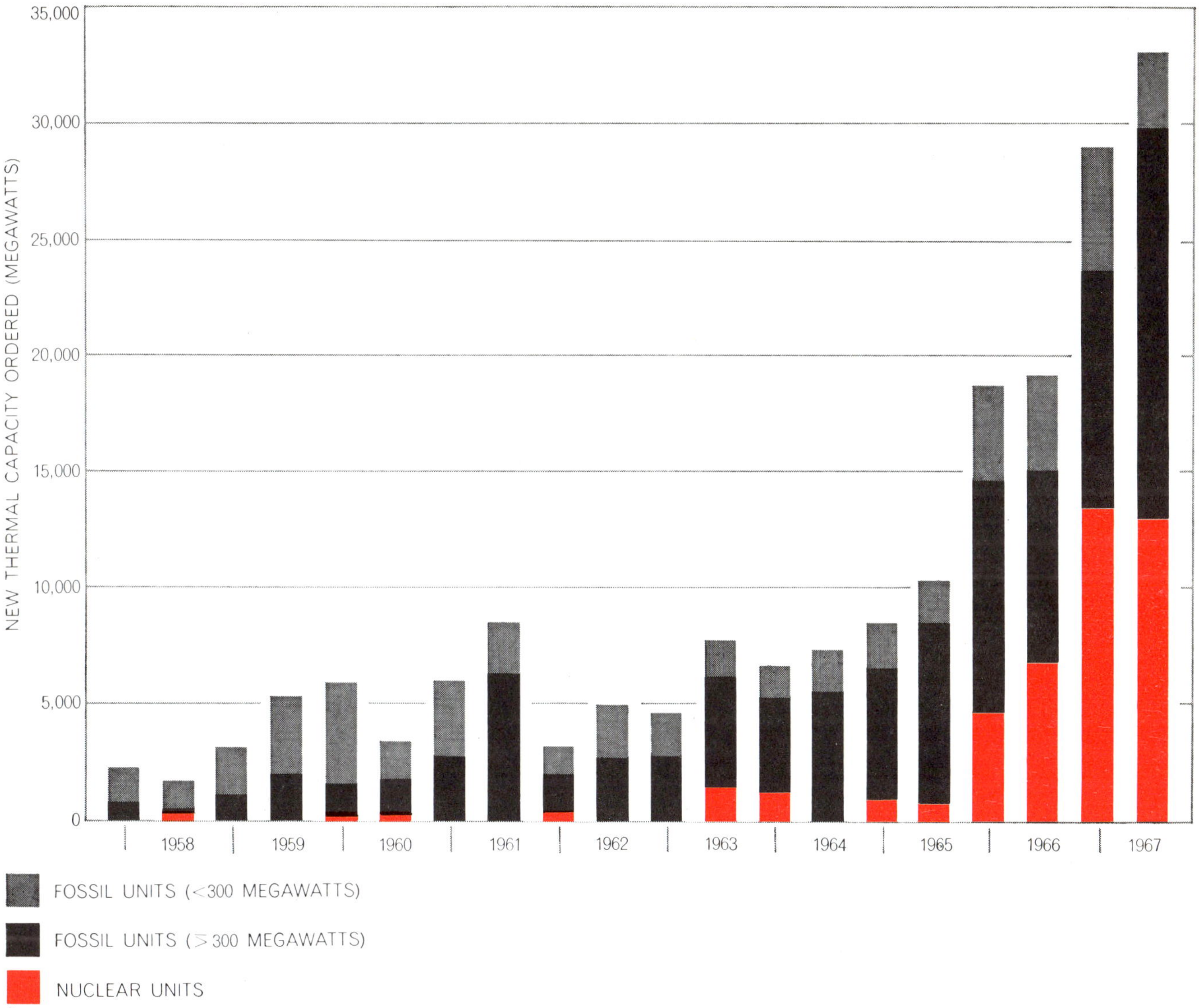

35,000
30,000
25,000
20,000
15,000
10,000
5,000
0
NEW THERMAL CAPACITY ORDERED (MEGAWATTS)
1958
1959
1960
1961
1962
1963
1964
1965
1966
1967
FOSSIL UNITS (<300 MEGAWATTS)
FOSSIL UNITS (≥300 MEGAWATTS)
NUCLEAR UNITS

2

3

Million short tons
0
1
2
3
4
UNITED STATES
INDIA
U.S.S.R.
MAINLAND CHINA
NIGERIA
PHILIPPINES
INDONESIA
ARGENTINA
SPAIN
BRAZIL
Soybean
Cottonseed
Peanut
Coconut
Sunflower
Olive
Rapeseed
Palm and Palm Kernel
Linseed
Other
+ Chiefly castor, babassu, cottonseed and peanut oil.

4

Africa
Asia
North America
Europe
South America
Oceania

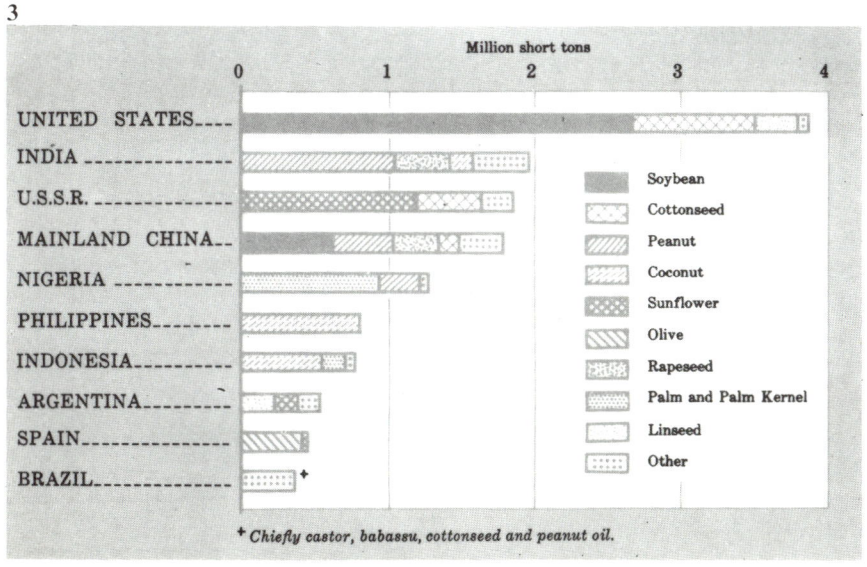

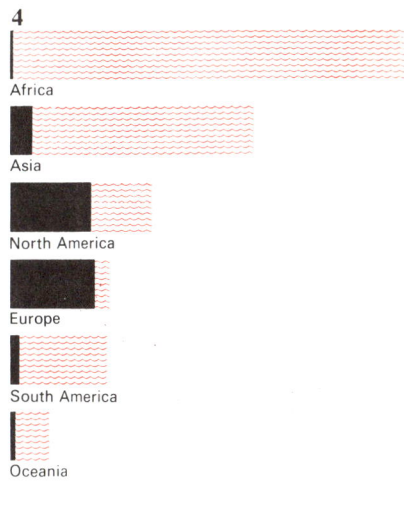

INCREASE IN AIR PASSENGERS

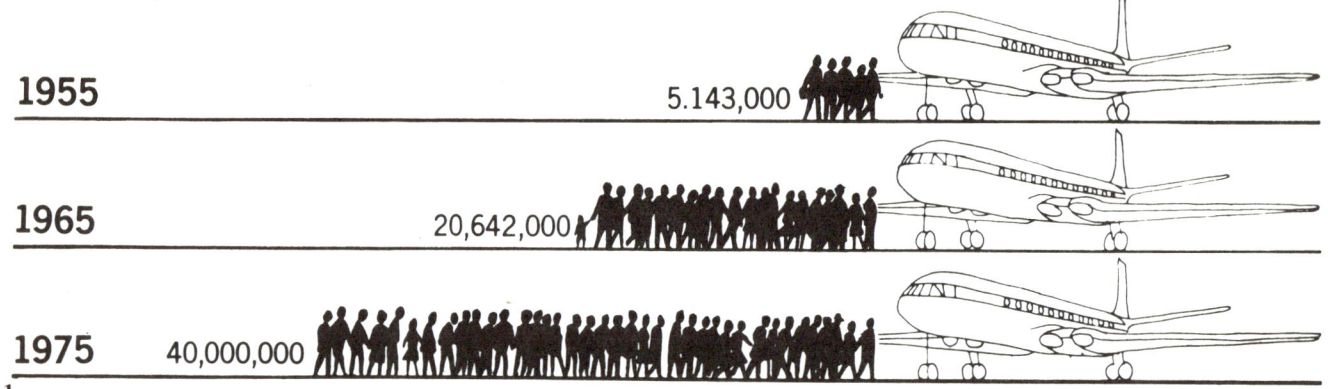

1955
5.143,000
1965
20,642,000
1975
40,000,000

1

2

U. S. TARIFF HISTORY

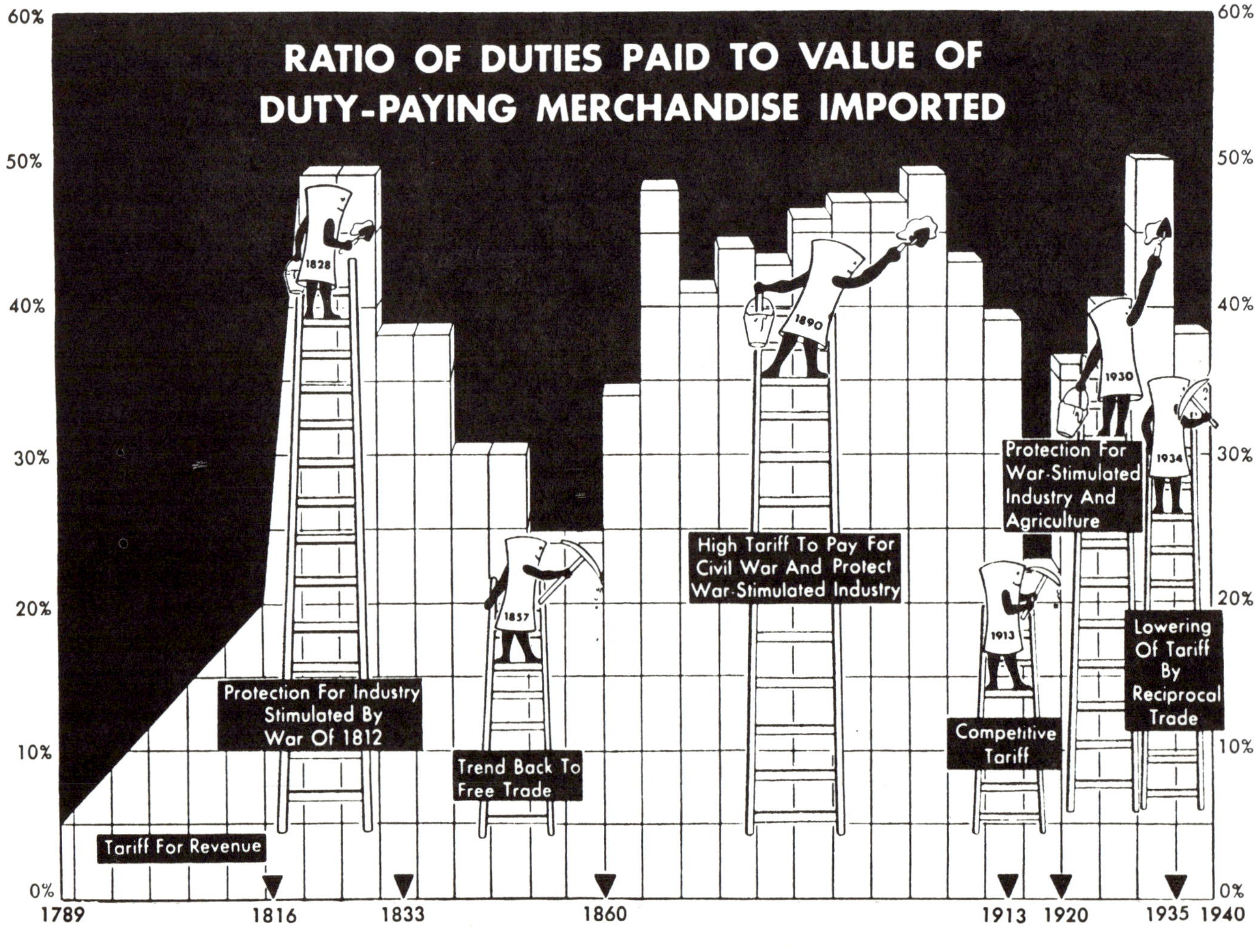

RATIO OF DUTIES PAID TO VALUE OF DUTY-PAYING MERCHANDISE IMPORTED
60%
50%
40%
30%
20%
10%
0%
1828
1857
1890
1913
1930
1934
Tariff For Revenue
Protection For Industry Stimulated By War Of 1812
Trend Back To Free Trade
High Tariff To Pay For Civil War And Protect War-Stimulated Industry
Competitive Tariff
Protection For War-Stimulated Industry And Agriculture
Lowering Of Tariff By Reciprocal Trade
1789
1816
1833
1860
1913
1920
1935
1940

Bar graph

The great problem the designer faces with graphs is giving them an appearance which will attract and interest (line graph solution page 13). This is particularly true at certain levels, especially in newspapers and school text books. Ingenious solutions are often devised – many must appal statisticians but they make for memorable images.

1 Bar graph from a school publication, showing the increase in the number of aircraft passengers. The length of the line of the passengers represents the number carried; it is not a pictorial graph (see page 56). BBC Publications

2 Graph from a school atlas, showing ratio of duty paid to value of duty-paying merchandise imported. *American History Atlas,* Hammond

3 Bar graph from a newspaper, showing car production. Separating the bars and superimposing them on an illustration overcomes the 'dead' quality of the bar graph. Clear labelling overcomes the faults which result from this approach. *Weltwoche*

The aim of the Nuffield Project is to devise a 'contemporary approach for children from 5 to 13'. The result is a revolution in visual representation of mathematical problems and exciting use by children of diagrams to sort out mathematical data. Books produced for the Nuffield Project are illustrated where possible from diagrams produced by the children.

4 Graph from a Nuffield Project text book, showing the village where children from one class live. *Pictorial Representation,* W. & R. Chambers and John Murray

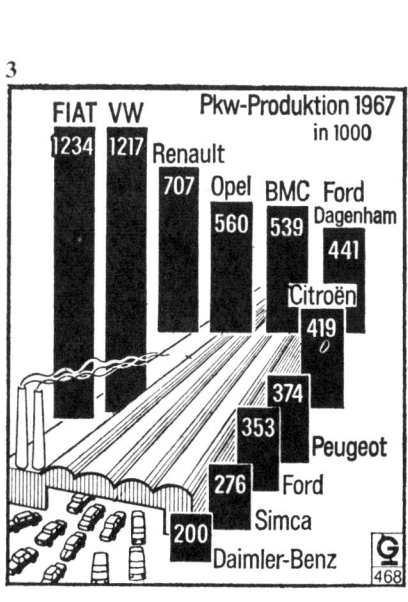

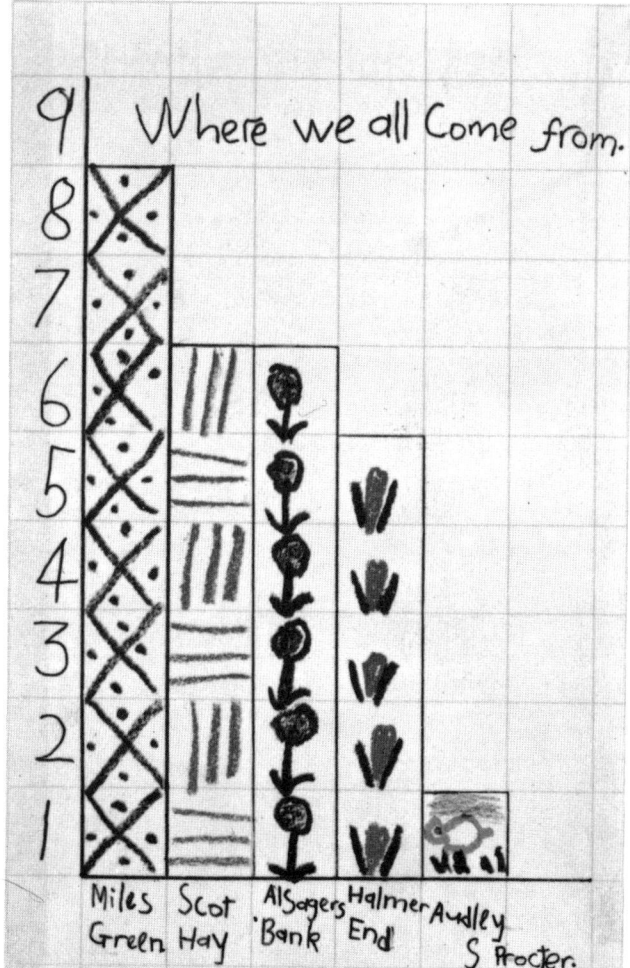

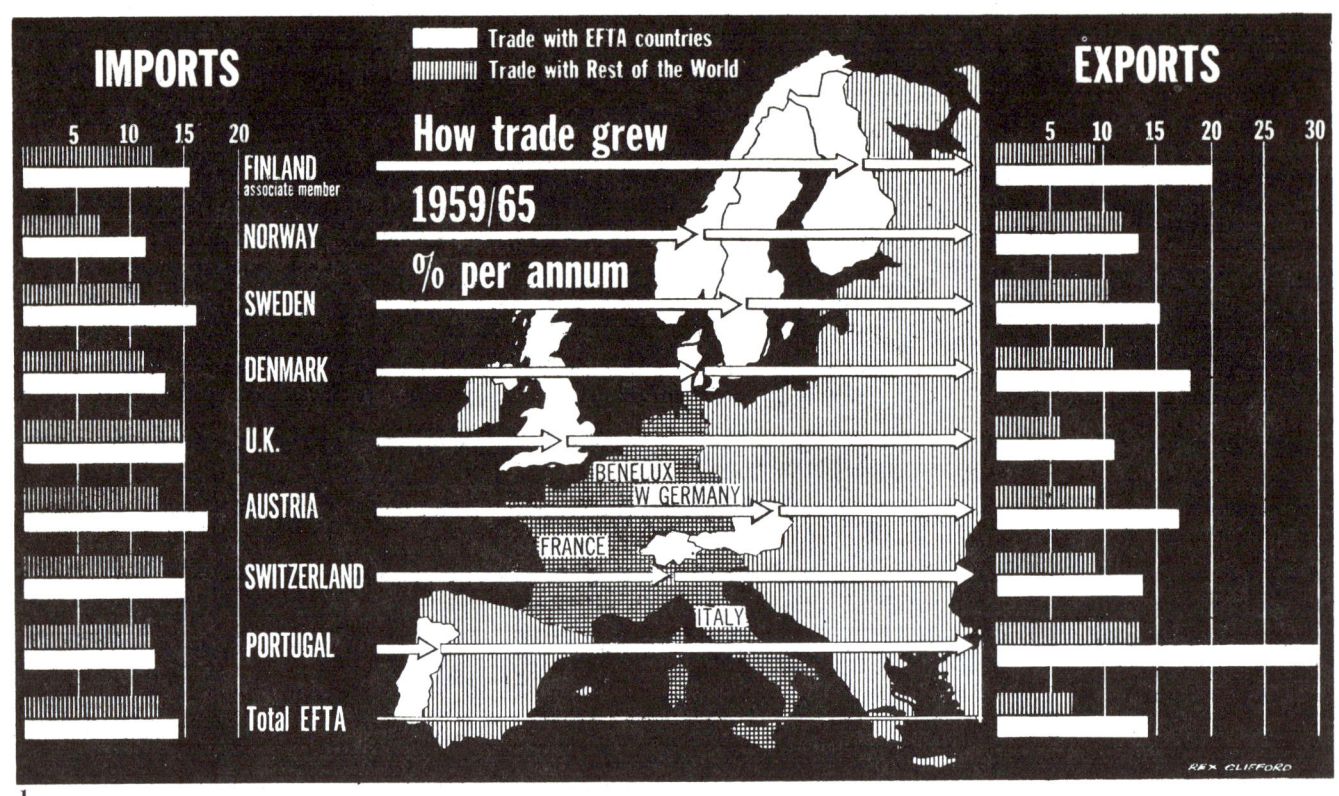

1

2

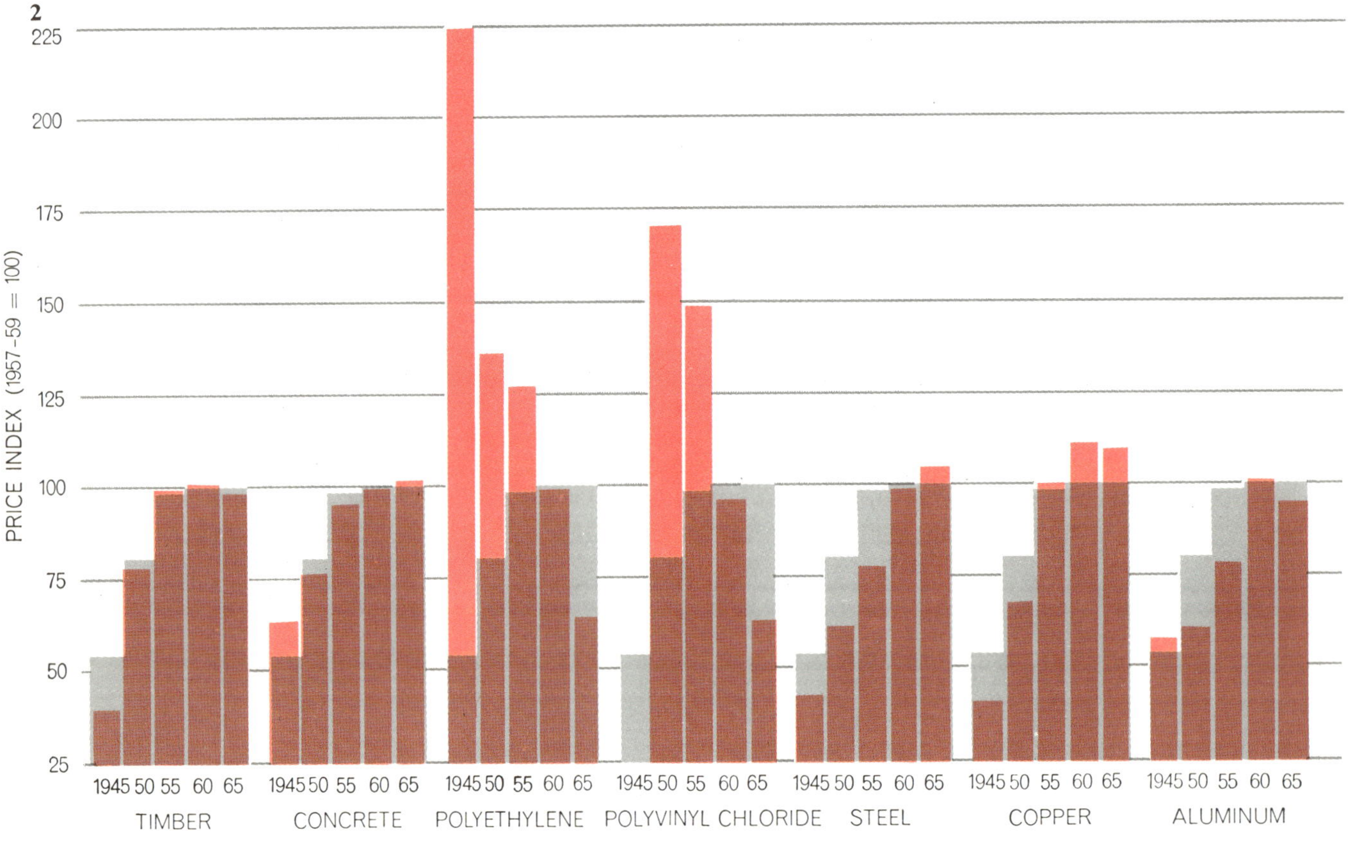

Bar graph

1 Graph from a newspaper, showing imports and exports to EFTA countries. Two bars side by side placed horizontally – are they easier to read than bars placed vertically? *Sunday Times*

2 Graph from a scientific magazine, comparing prices of various materials with a price index in the US. Two bars, colour coded, completely overlap. Pink and grey can be easily read, and the key information sorted out, e.g. polyethylene, priced 225 in 1945, dropped to 62 in 1965. *Scientific American*

3 Graph from a newspaper, showing unfilled vacancies and the number of wholly unemployed. Bars overlap. *Sunday Times*

4 Graph from a paperback history atlas showing industrial production in pounds sterling of Russia, Germany, France and Great Britain. Four bars overlap. Original in colour. *dtv-Atlas zur Weltgeschichte,* Deutscher Taschenbuch Verlag

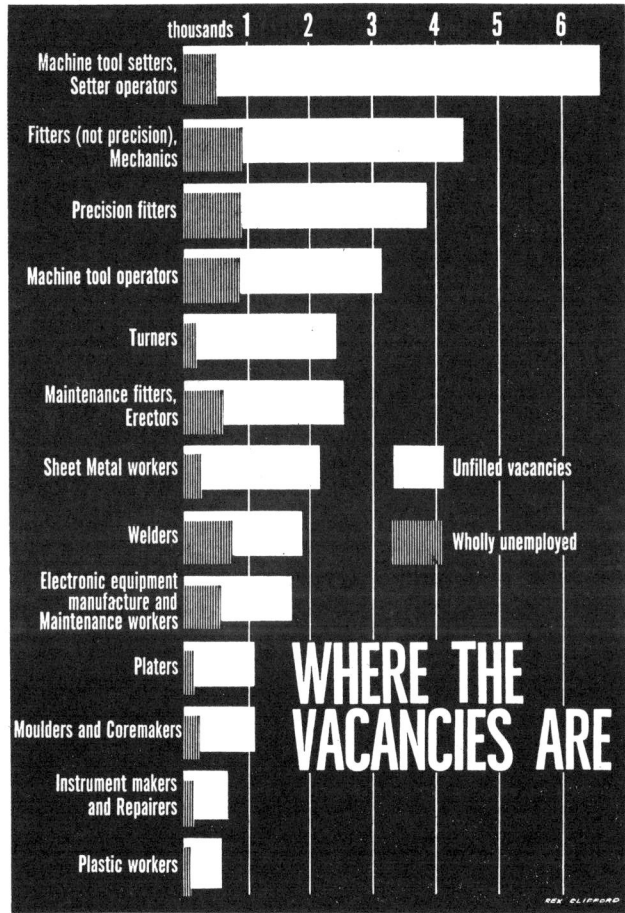

Floating bar graph

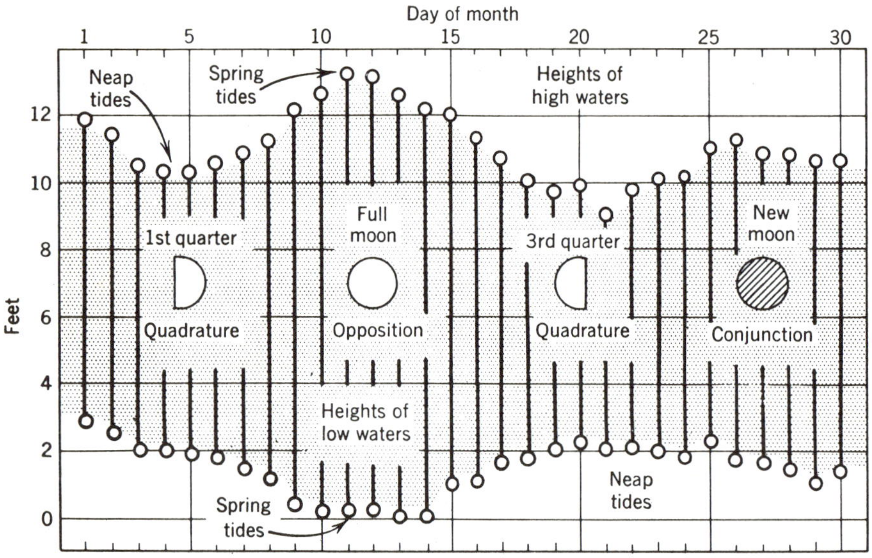

The bar 'floats', either in the area of the graph, or above and below a zero line from which a value scale runs down as well as up.

1 Graph from a text book, showing neap and spring tides. The line 'floats' in the graph and represents the daily height of high and low tide over a period of a month. A. N. Strahler, *Physical Geography,* John Wiley

2 Graph from a newspaper, showing share price range for a company. *Sunday Times*

3 Graphs from a text book, showing the highest water level that occured each month in terms of percentages for two rivers. A geographer's diagram. A. N. Strahler, *Physical Geography,* John Wiley

4 Graph from a scientific magazine, showing proved world reserves of crude oil. Upper segment of bar shows production, lower segment of bar shows proved reserves. Colour coding indicates that 1950 reserves have been surpassed by subsequent production, but in the meantime the reserves have more than tripled. *Scientific American*

5 Graph from a magazine showing urban development. The left hand shapes show the population of selected urban districts. Figures for 1860 are at the bottom, those for 1960 at the top. The right hand shapes show the percentage of the national population living in these districts in the same years. *History of the 20th Century,* Purnell

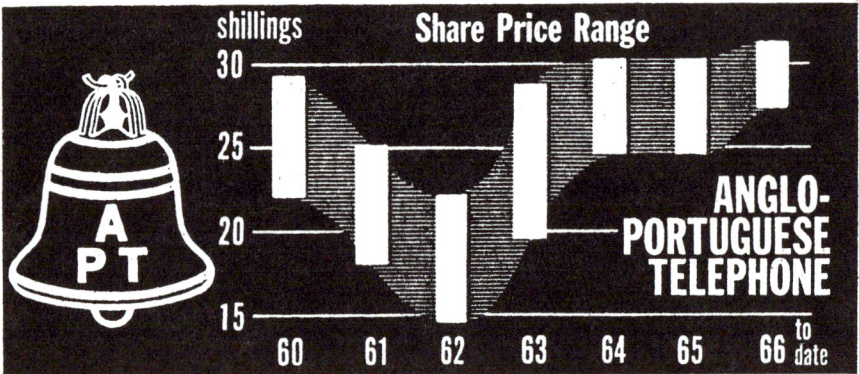

1

2

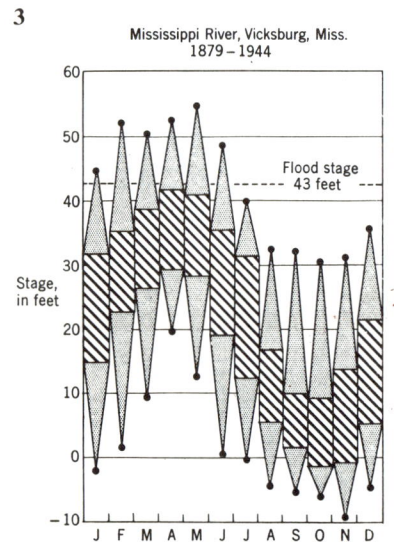

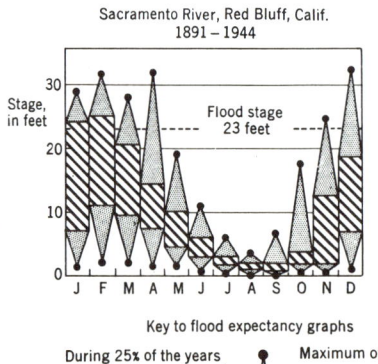

3

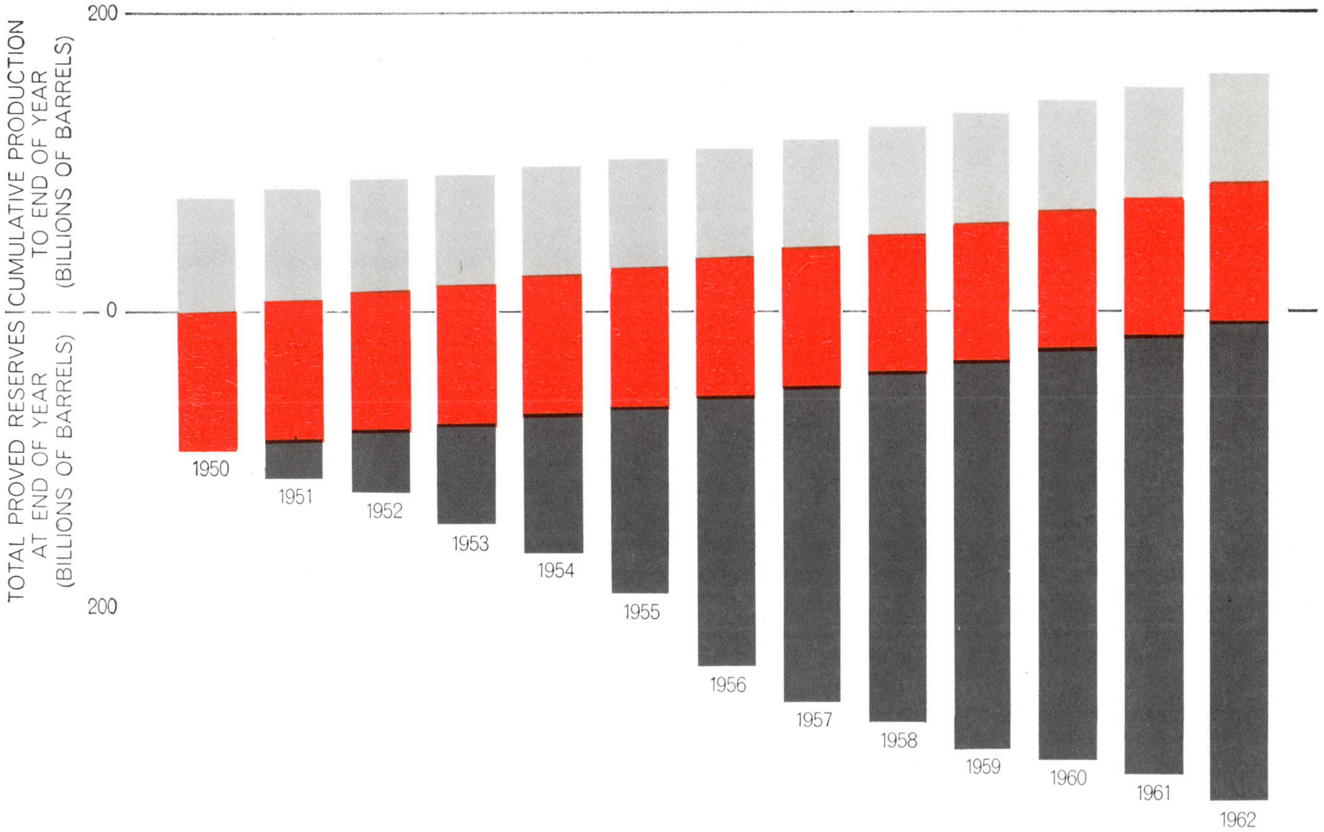

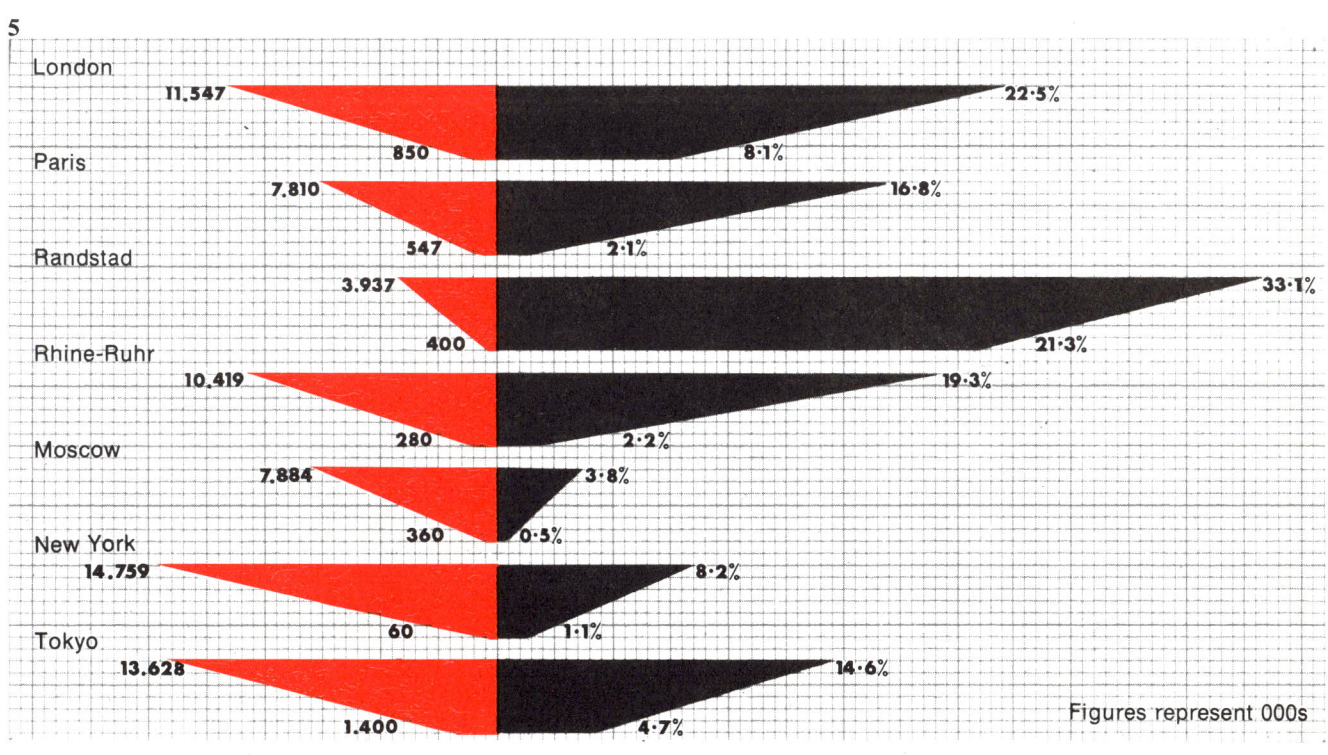

Population pyramid

The method of showing the sex and age structure of a population is to represent it in the form of a pyramid built up in age groups, male on one side, female on the other. The ages can be grouped every one, five or ten years and the information can be read, apart from the length of the bar, by comparing the shape of one pyramid with another.

1 Population pyramid from a scientific magazine, showing US figures for 1950 (dark), compared with 1960 (light). The small unshaded area in age group '20–29' shows population loss since 1950. Clear grid enables information to be read easily. Five-year groups. Reduced from 7". *Scientific American*

2 Population pyramids from atlas, showing figures for selected countries. This diagram has ten-year groups, and illustrates clearly in simple form the differences in age structure produced by historical and economic factors. Ghana and Ceylon are undergoing an explosion of population as a result of declining infant mortality. In the UK and West Germany the marks of two wars show clearly – the low birth rate and the increasing number of elderly dependents. *Shorter Oxford Economic Atlas,* Oxford University Press

3 Population pyramids from a paperback encyclopedia, showing figures for Germany 1910, 1939, 1959. One-year groups. (Detailed information for one-year groups can be difficult to get, but quinquennial data are published in international returns. *dtv-Lexikon,* Deutscher Taschenbuch Verlag

4 Population pyramid from a school text book, showing figures for England and Wales. Pictorial representation is used to 'humanize' diagram. Marie Neurath, *Living with One Another,* Parrish

5 Divided pyramid from a text book showing total population for each census year, proportional in length. It is divided into rural and urban parts. F. J. Monkhouse and H. R. Wilkinson, *Maps and Diagrams,* Methuen.

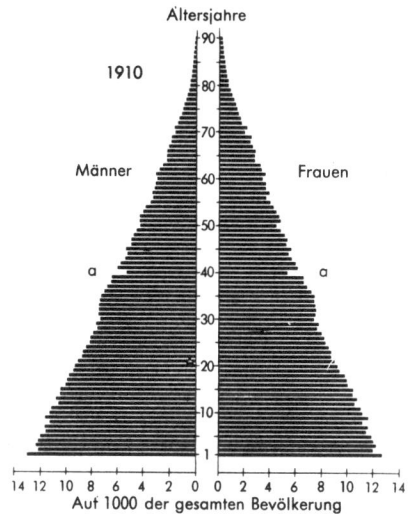

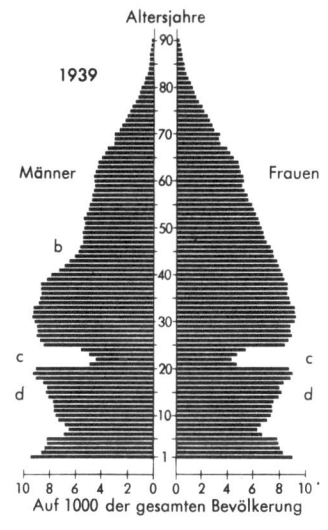

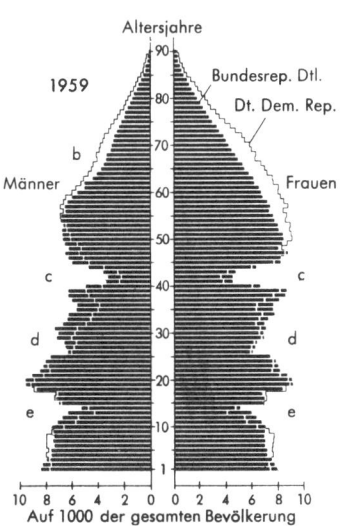

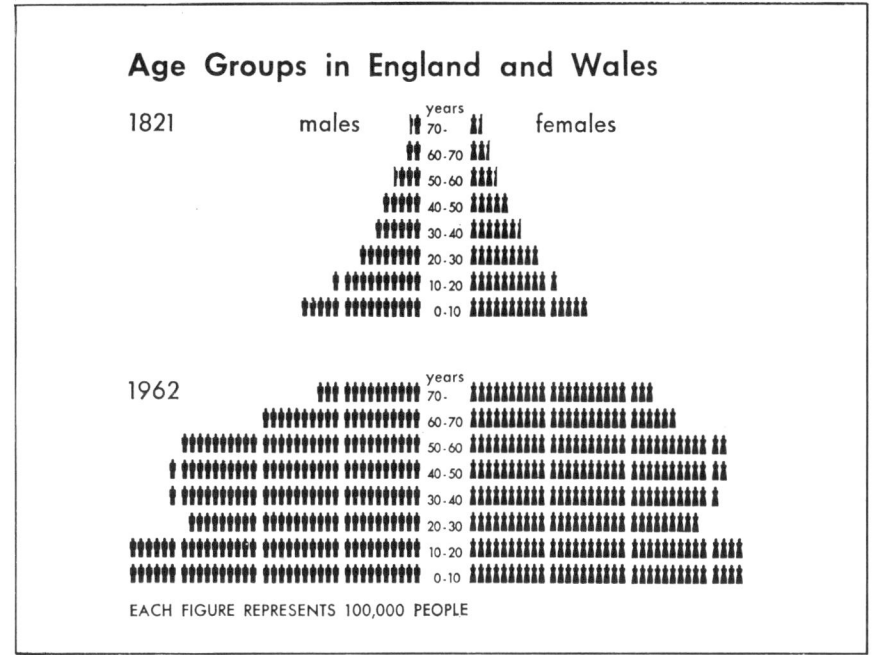

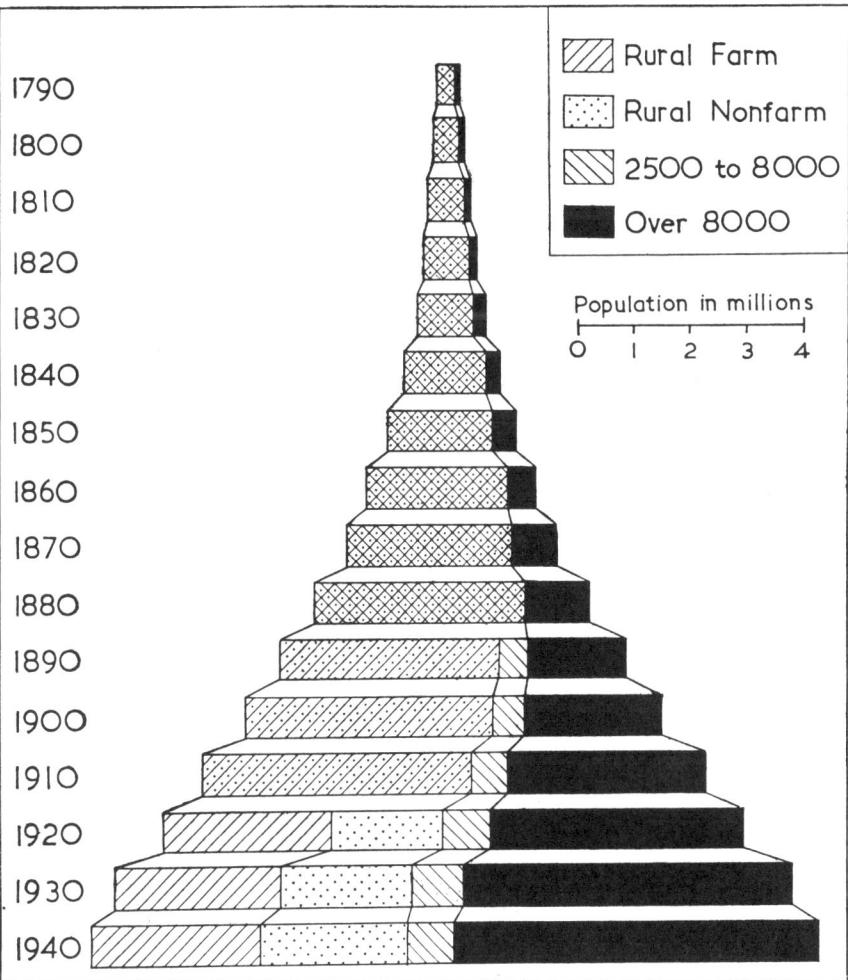

1 Histogram from a scientific magazine, showing the frequency at which eye movements occur during sleep. It records in percentage form eye movements during a sequence of nights. *Scientific American*

2 Graph from a business publication, showing tonnage sales of chocolate compared with price per ½ lb. of chocolate. The bar has been dropped to achieve better visual comparison between the two elements. There are two scales. *Industrial Challenge*, Cadbury Brothers

INCIDENCE OF EYE MOVEMENT
(PER CENT)
75
50
25
0
1
2
3
4
5
6
7
HOURS

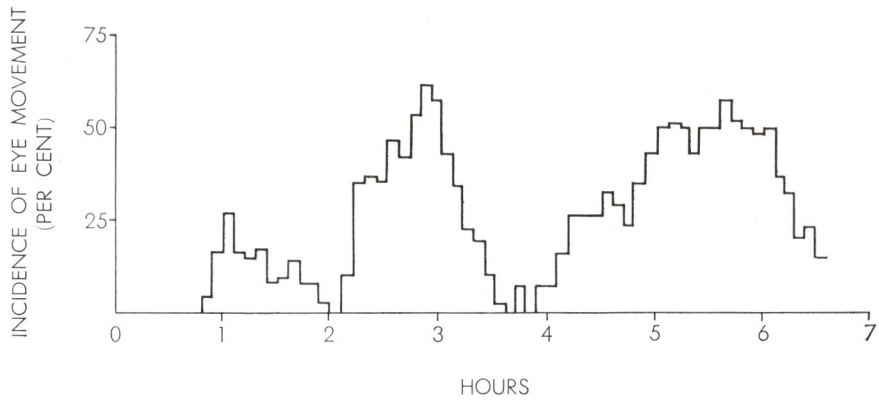

1

2

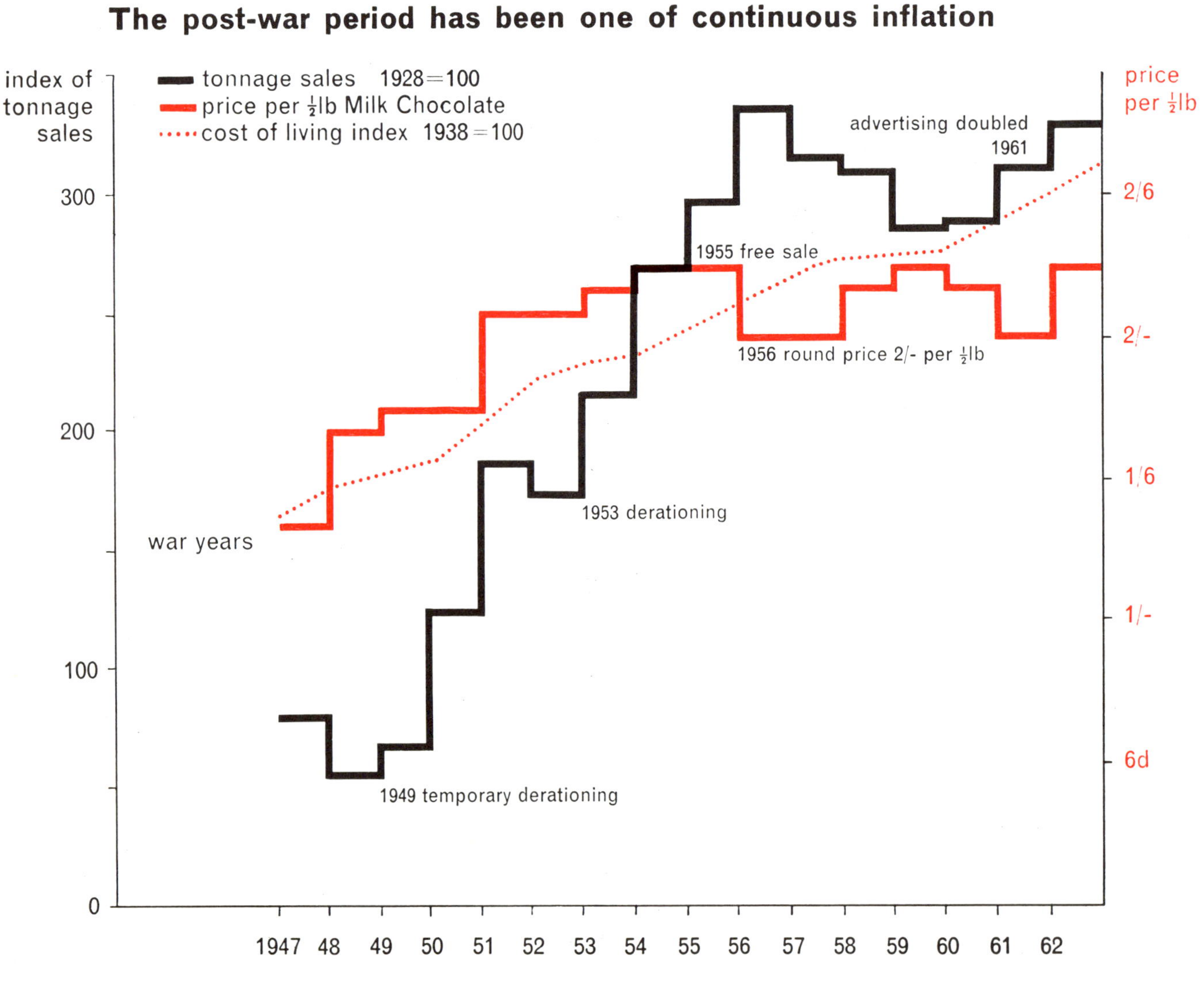

The post-war period has been one of continuous inflation
index of tonnage sales
tonnage sales 1928=100
price per ½lb Milk Chocolate
cost of living index 1938=100
price per ½lb
advertising doubled 1961
300
2/6
1955 free sale
2/-
1956 round price 2/- per ½lb
200
1/6
1953 derationing
war years
1/-
100
6d
1949 temporary derationing
0
1947
48
49
50
51
52
53
54
55
56
57
58
59
60
61
62

Bar line graph

3 Diagrams from a school text book, giving the same information using a graph and a simple bar line graph. The exact point for each hour (or half or quarter hour) can be clearly shown *and* read. Raymond S. Heritage, *Learning Mathematics,* Penguin Education

4 Graph from a scientific magazine, showing competition between nuclear power and fossil power, related to cost. A complicated diagram with a lengthy caption and good key. Lines enable each power station to be shown, and the length of bar at the end of the line gives additional information. *Scientific American*

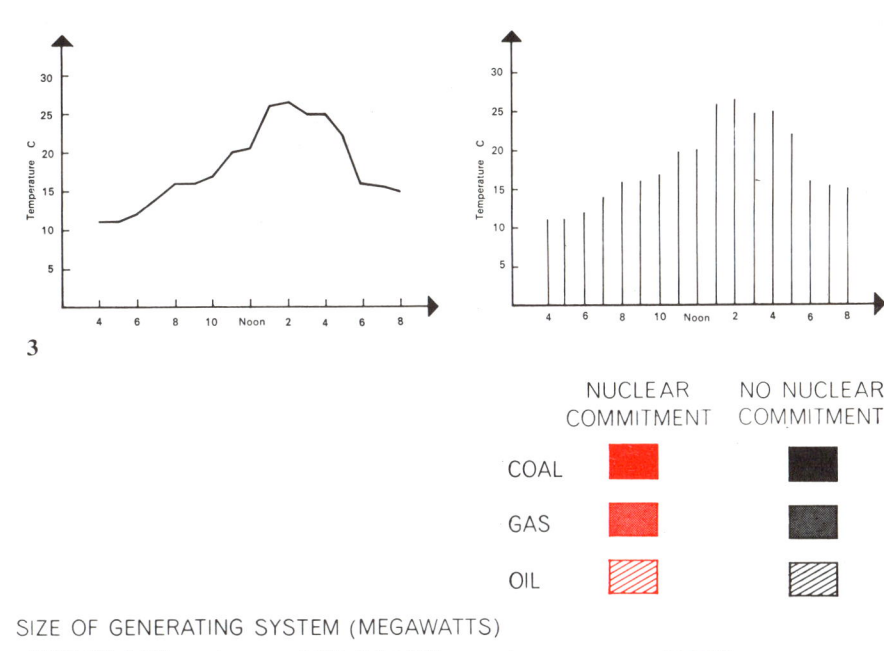

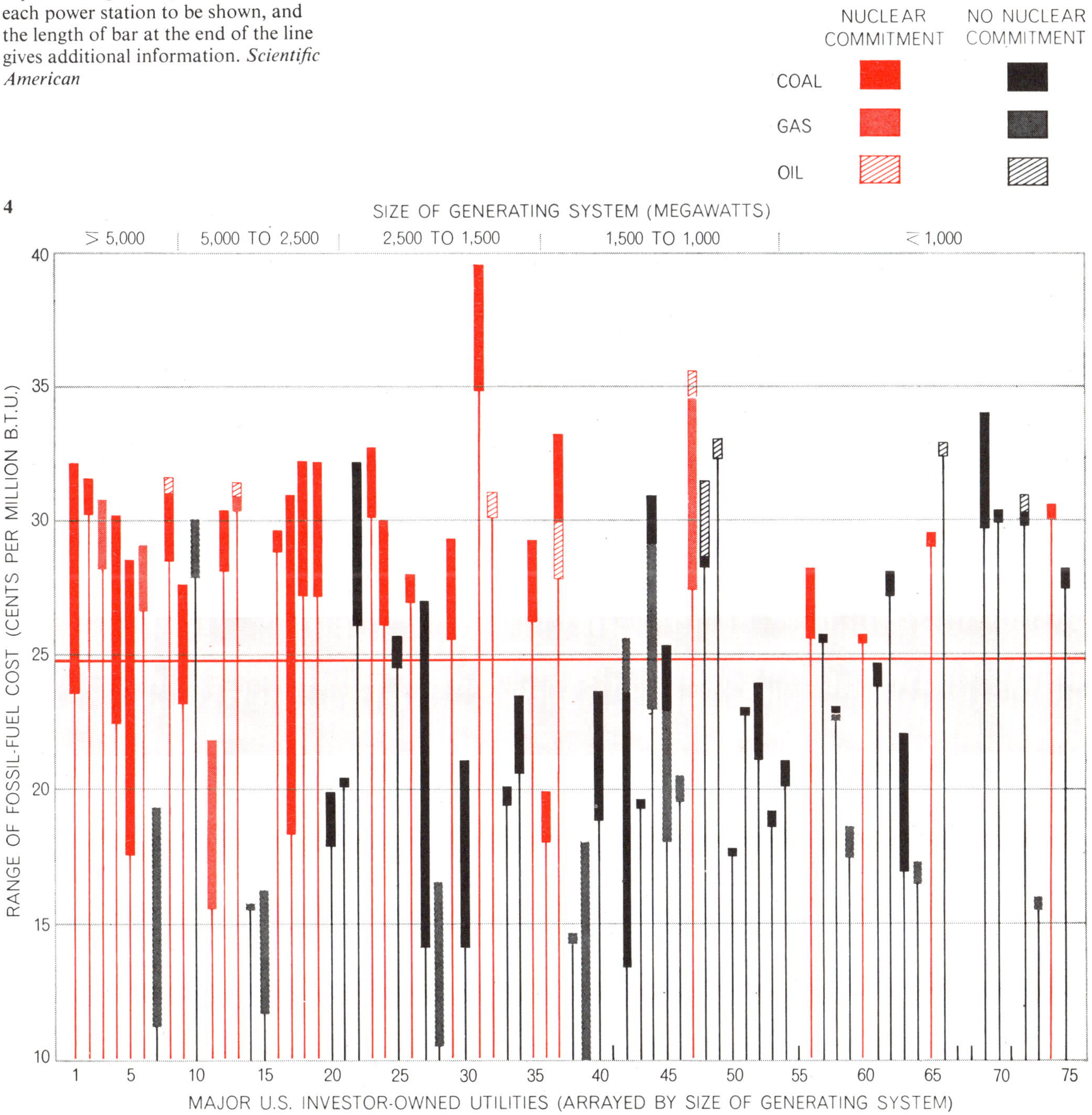

Bar diagram

The exact difference between the bar graph and bar diagram is not always clear. Bar graphs merge into diagrams which merge into block diagrams. Classification is often a matter for easy reference, but whereas a bar graph may be interchanged with the line graph, the bar diagram could never be designed as a line graph.

Each item is clearly separate; comparison is made by bars usually of the same width which represent the information. Since there are no graph lines to give figures, labelling is placed in the bar or at the end.

1 Advertisement for motor oil. An adapted bar graph with everything omitted except the bar and the total figure. Final bar separated and solid blank; other bars tones of grey. A diagram is not always the most convincing way to show superiority of a product. Esso

2 Bar diagram from a scientific magazine, showing how lack of education has handicapped negroes. The diagrams compare segregated public school services for whites (grey) and negroes (black) for certain States in certain pre-Second World War years. Each statement is simple with no common measurement (percentages, days, pupils per teacher, dollars). It would not be possible to use all these on one bar graph. *Scientific American*

3 Diagram from an educational book showing the distance travelled per second by various insects. Arrowhead introduces variable on simple bars and suggests nature of diagram – movement. L. Hugh Newman, *Man and Insects*, Aldus Books

4 Diagram from a magazine showing acceleration (how many seconds taken to reach 60 mph) and braking (how many feet taken to stop from 60 mph). It is impossible to compare bars of different length without measuring, if they do not start from the same point. *Sunday Times*

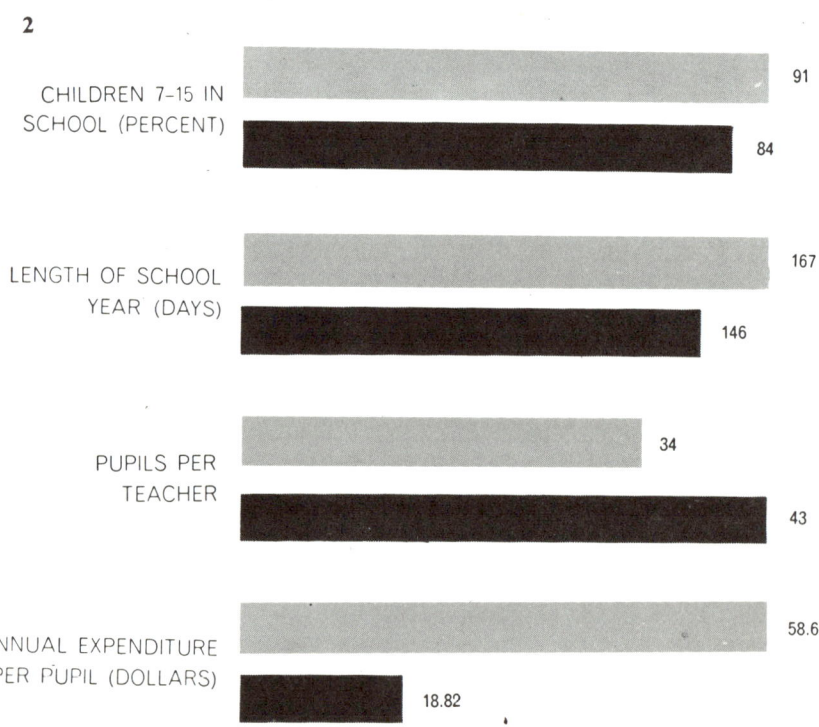

	wing beats per second		distance travelled per second
Hornet	100		
Hive-bee	250		
Large White Butterfly	12		
Housefly	190		
Humming-bird Hawk Moth	85		
Bumblebee	130		
Cockchafer	46		
Aeshna Dragonfly	38		

3

4

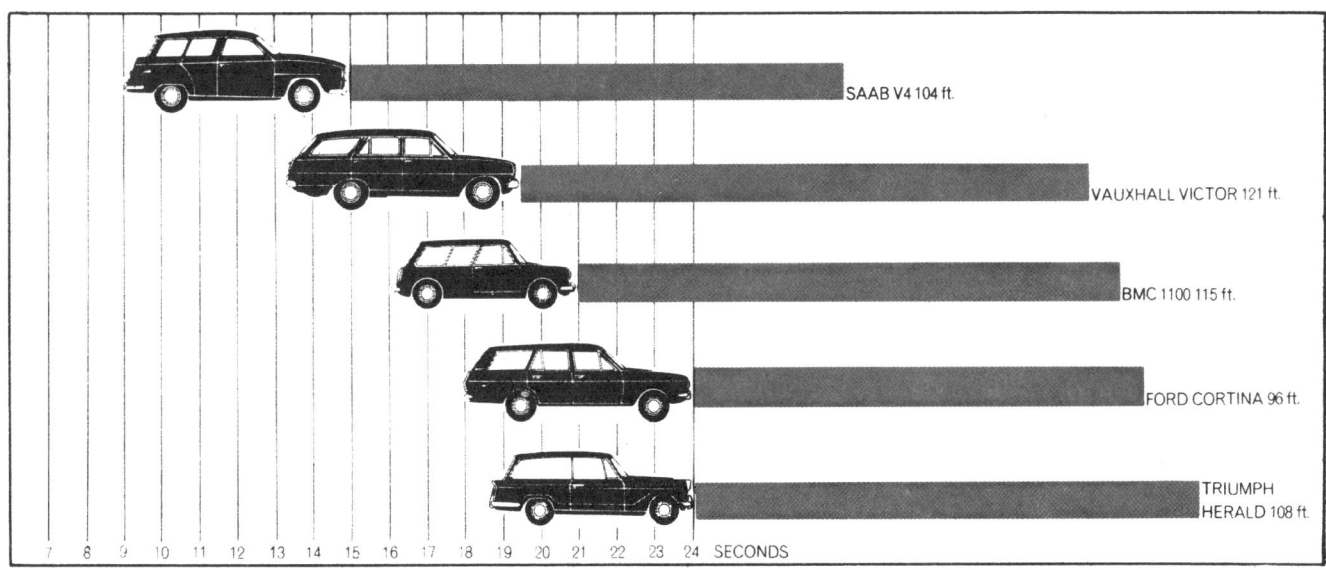

SAAB V4 104 ft.
VAUXHALL VICTOR 121 ft.
BMC 1100 115 ft.
FORD CORTINA 96 ft.
TRIUMPH HERALD 108 ft.

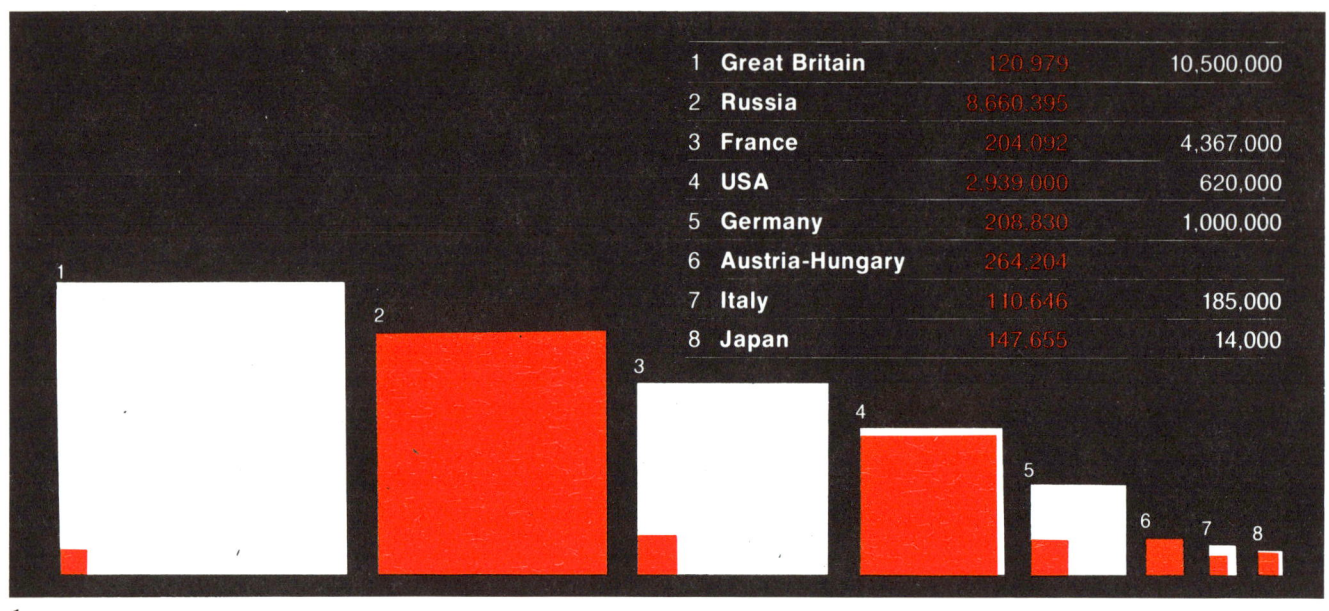

1

1	Great Britain	120,979	10,500,000
2	Russia	8,660,395	
3	France	204,092	4,367,000
4	USA	2,939,000	620,000
5	Germany	208,830	1,000,000
6	Austria-Hungary	264,204	
7	Italy	110,646	185,000
8	Japan	147,655	14,000

2

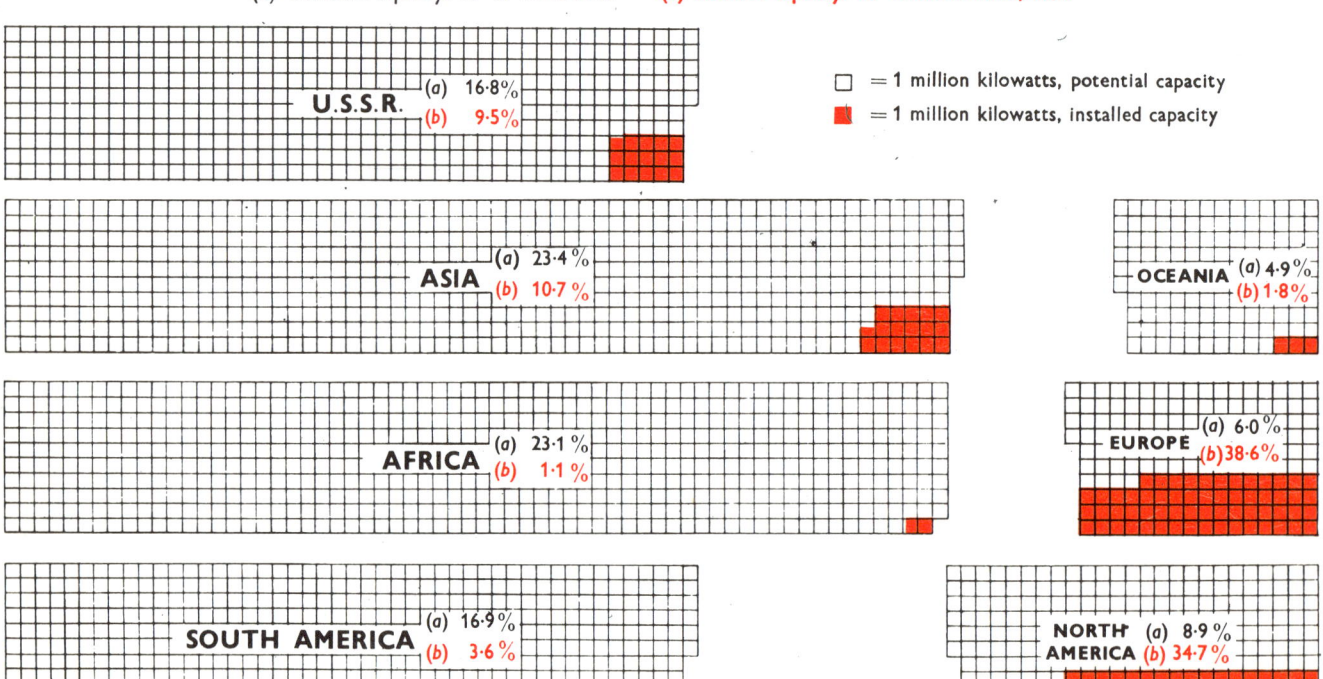

Potential and Developed Water Power
Total world installed capacity 1960 154·6 million kilowatts
Total world potential capacity, at mean flow, 2,762 million kilowatts.

(a) Potential capacity: % of world total   (b) Installed capacity: % of world total, 1954

□ = 1 million kilowatts, potential capacity
■ = 1 million kilowatts, installed capacity

U.S.S.R. (a) 16·8% (b) 9·5%

ASIA (a) 23·4% (b) 10·7%

OCEANIA (a) 4·9% (b) 1·8%

AFRICA (a) 23·1% (b) 1·1%

EUROPE (a) 6·0% (b) 38·6%

SOUTH AMERICA (a) 16·9% (b) 3·6%

NORTH AMERICA (a) 8·9% (b) 34·7%

Block diagram

Figures being compared are represented by a 'block' or square. Sometimes the square is only the basic unit and is used to form a larger square or possibly a rectangle. But once the rectangle is introduced, the diagram becomes nearer to the bar graph or bar diagram and loses what advantage the 'block' has. For simple sets of figures the block diagram is useful, but if the information is complex the bar graph will probably be the better method to use. It is much easier to compare the length of bars which have the same width than it is to compare the sizes of squares or cubes which vary in height and width.

The block diagram can be used to advantage in showing information related to its nature, i.e. figures dealing with area or volume. The square suggests area and comparison of squares of various size can have impact.

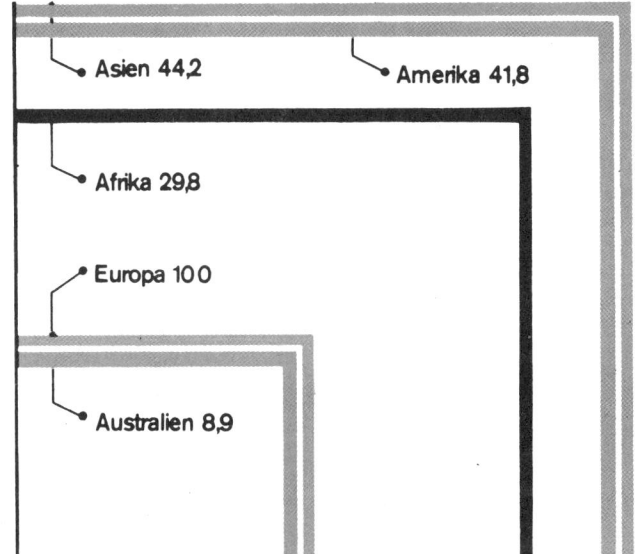

3

1   Diagram from a magazine, showing size in square miles of states related to their overseas empires. Figures are included to overcome the problem of comparing sizes of squares. *History of the 20th Century,* Purnell

2   Diagram from an atlas, showing potential and developed water power. The block is part of a larger unit. *Shorter Oxford Economic Atlas,* Oxford University Press

3   Diagram from a school text book, showing area of five continents in millions of square kilometres. By making each square have a common left hand side and base, the difficulties of easy comparison of squares of various size is overcome. A. Jentzsch and J. Winkler, *Güterkunde,* Westermann

4   Diagram from a book, showing the animal population, including man, of Africa. Comparison of sizes is not easy, but good labelling compensates, and symbols and labels give it a lively appearance. Georg Borgstrom, *Der Hungrige Planet,* Bayerischer Landwirtschaftsverlag

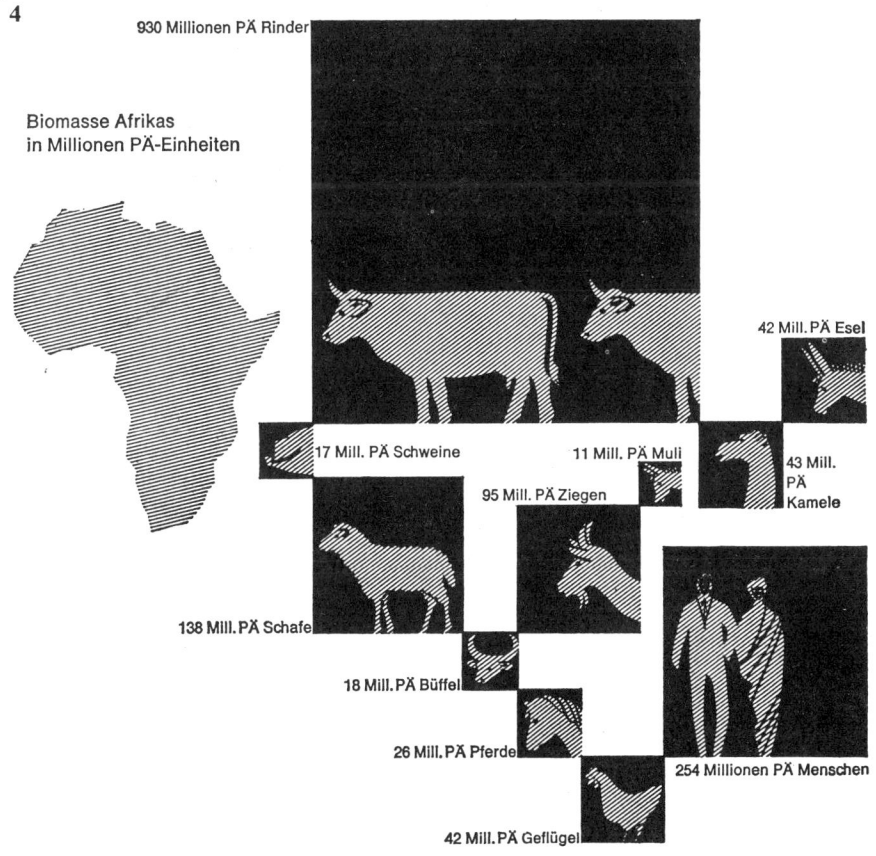

43

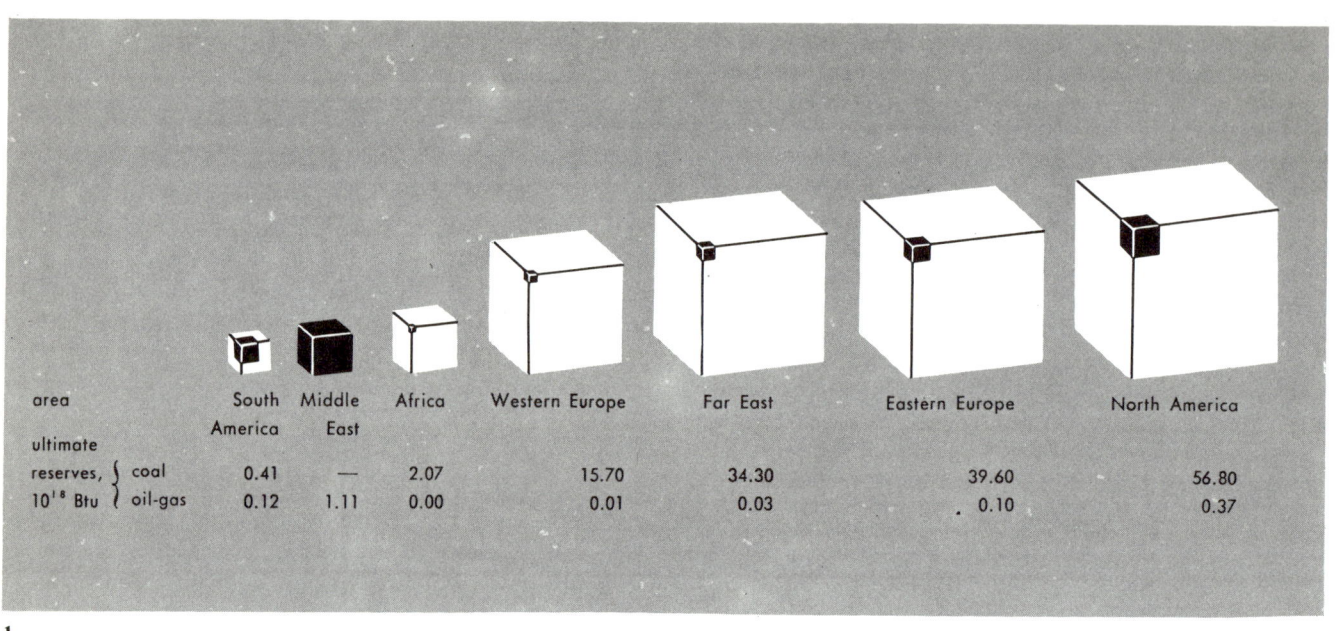

1

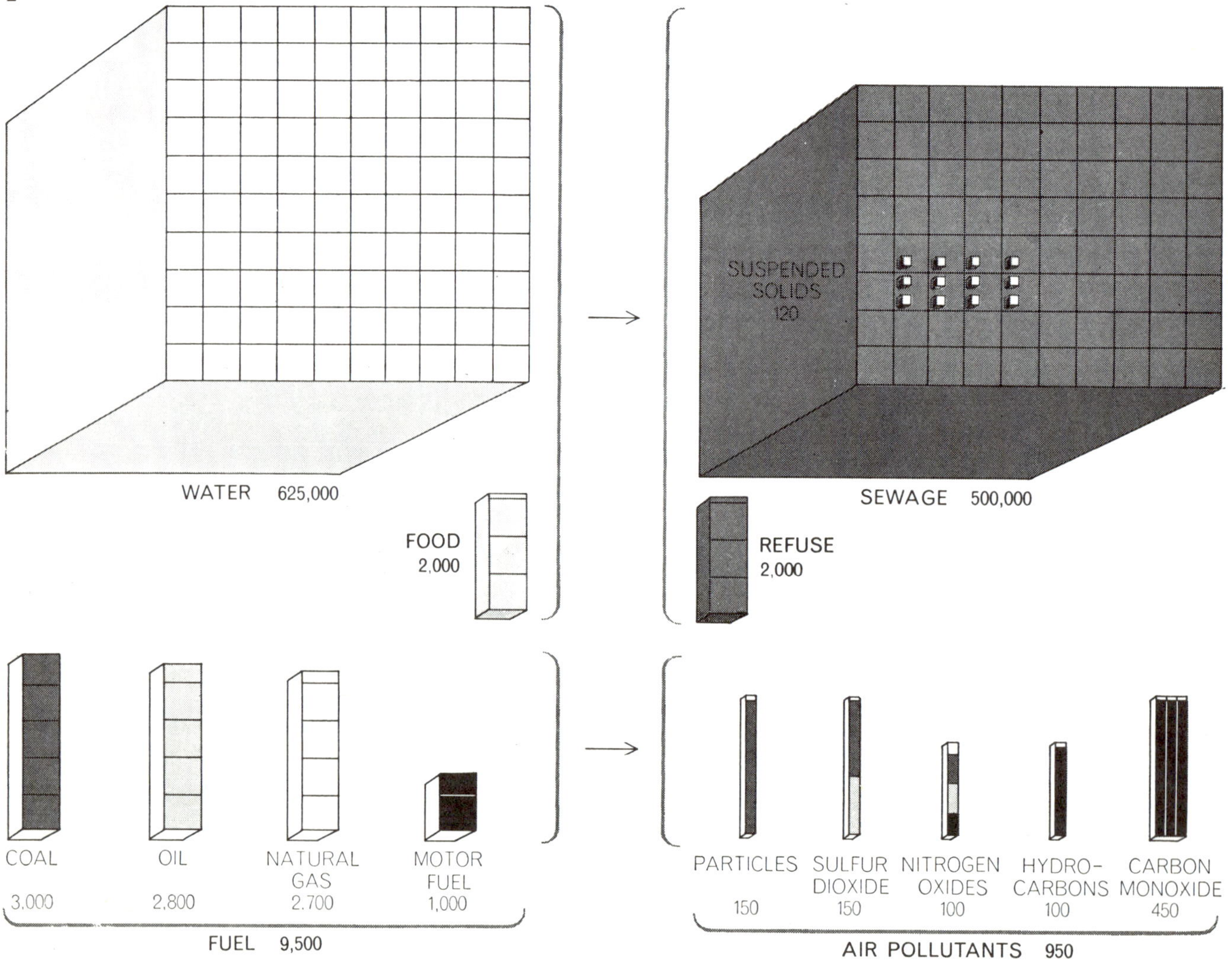

2

Block diagram

1   Diagram from a magazine, showing coal and oil-gas reserves in various areas. Comparison of a variety of sizes – this time of cubes. But what is the difference between South American and Eastern European oil-gas reserves? Figures are included to help with this problem. *Power*

2   Diagram from a scientific magazine showing input/output transactions of a city. On the left are three inputs: water, food and fuel, shown in tons per day for a US city of one million. On the right are outputs: sewage, refuse, and air pollutants. The use of cubes copes with the problem of an enormous range of numbers. *Scientific American*

3   Diagram from an annual report of a diversified company, showing the various parts of the company. Ingenious use of actual boxes of various sizes. Hunt Foods & Industries Inc.

4   Diagram from a school publication, showing growth of French trade. BBC Publications

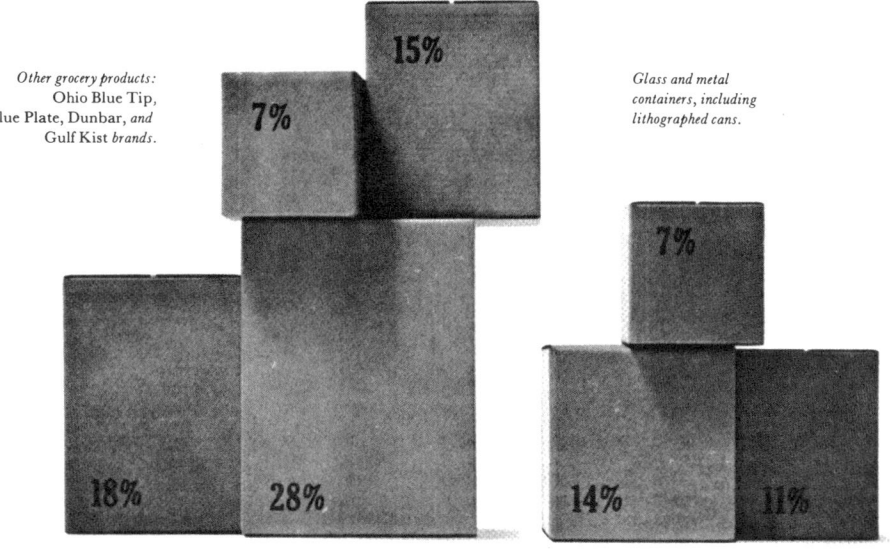

3

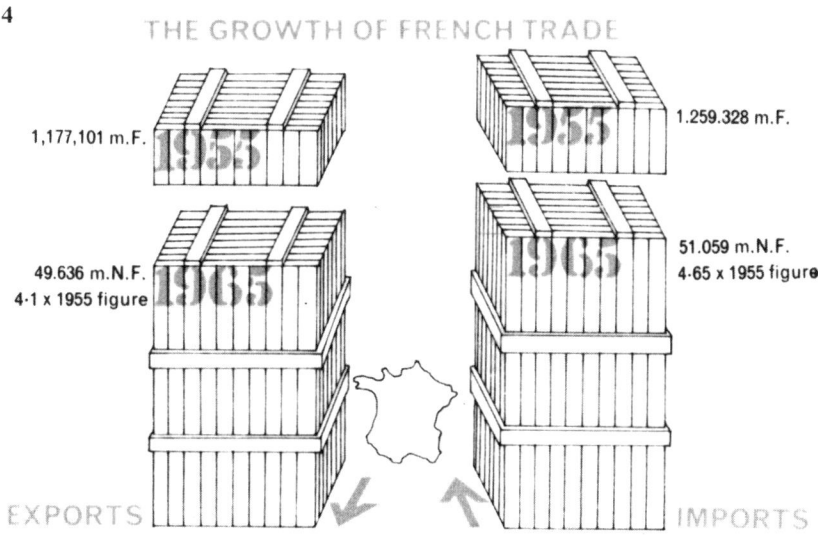

4

45

Divided rectangle

The simplest form is one rectangle divided into various parts. The most common use is the comparison of several rectangles of the same size divided on a percentage basis. Rectangles may sometimes vary in size as in a bar graph, but sizes and divisions will still be related on a percentage basis.

Comparison may also be concerned with area, and the area of the rectangle and of its divisions may be made proportional to other rectangles. This is always a difficult solution to follow, for one has to compare the size of rectangles of various proportions, and unless the percentage value is clearly shown it is difficult to read the information.

The divided rectangle has advantages over the divided circle (page 52) when comparisons have to be made with examples of differing size. The circle has only one dimension which can be varied, the radius. The rectangle has two, and variation can be got by changing only one of these and leaving the other the same.

1 Diagram from a scientific magazine, showing land use in seven metropolitan regions. Horizontal lines help to show the increase or decrease clearly.
Scientific American

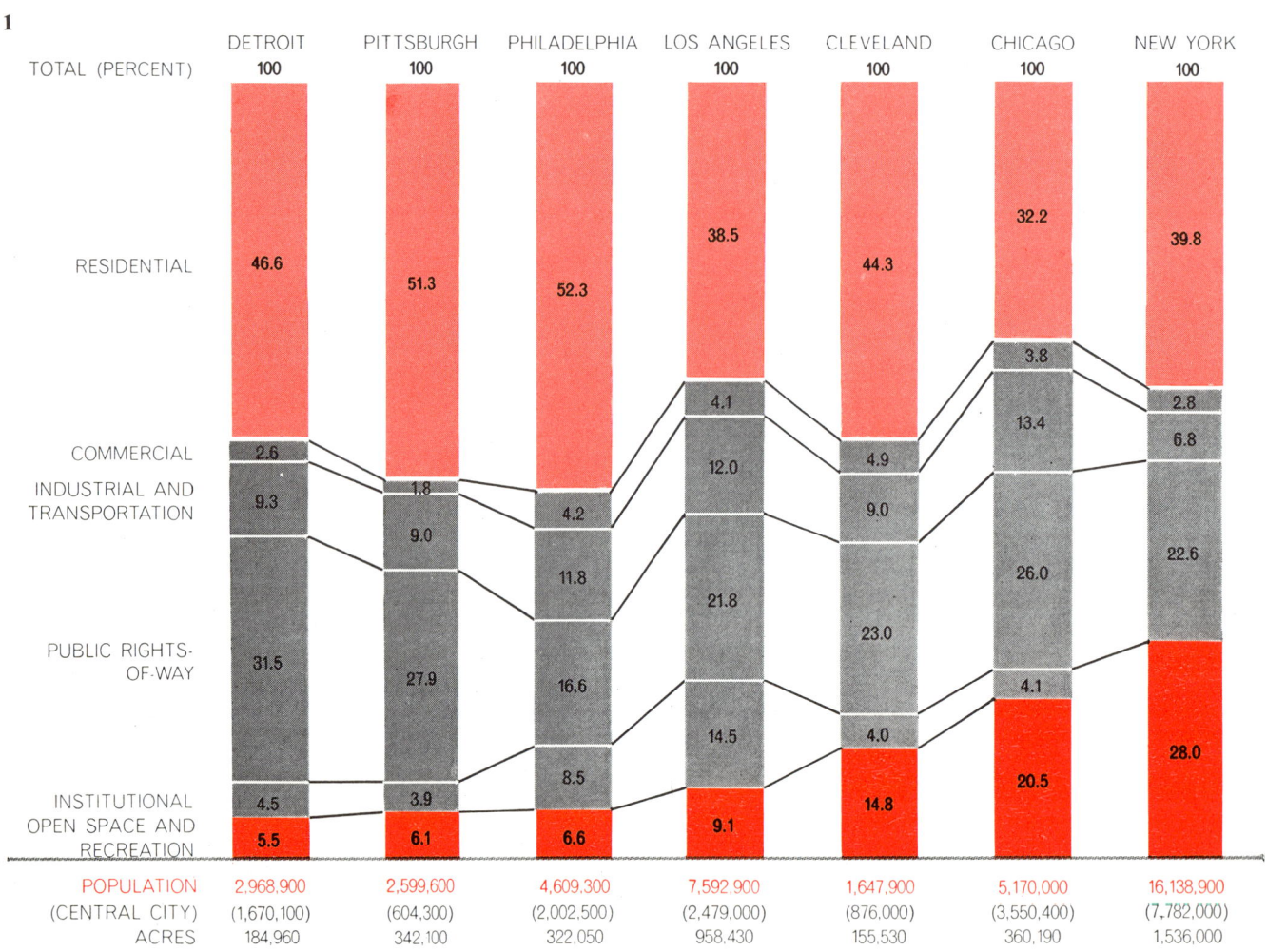

2 Diagram from an industrial publication, showing sources and uses of funds for one company, compared with all other quoted companies. The labelling problem is clearly organized, as are the textures chosen, so that the solid area shows the important elements, i.e. retained net income, fixed assets. *Industrial Challenge,* Cadbury Brothers

3 Diagram from an industrial publication, showing the cost of a bar of chocolate. The rectangle has been replaced by a 1938 and 1962 chocolate bar label. It has been divided in two ways: distribution and selling; materials and production. These are dealt with in more detail on the other side of the diagram so that the rise in cost of ingredients can be clearly shown. *Industrial Challenge,* Cadbury Brothers

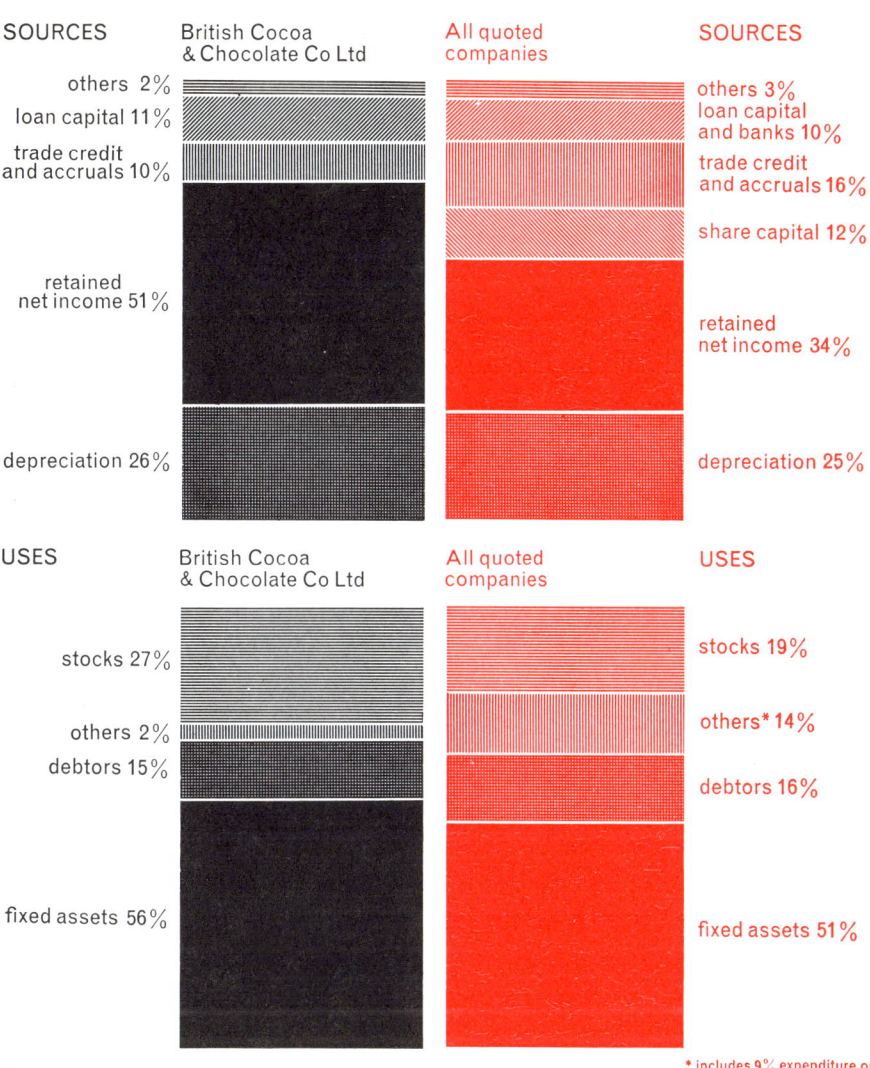

Tonnage sales have increased by 69 per cent since 1938

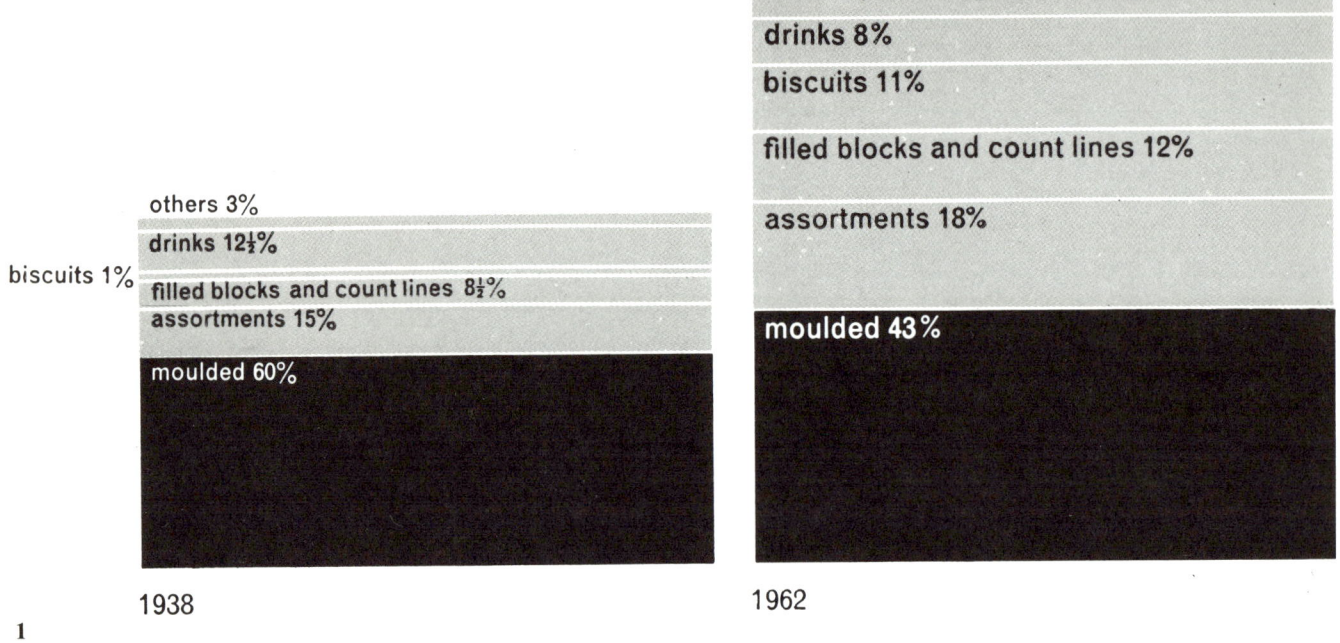

1

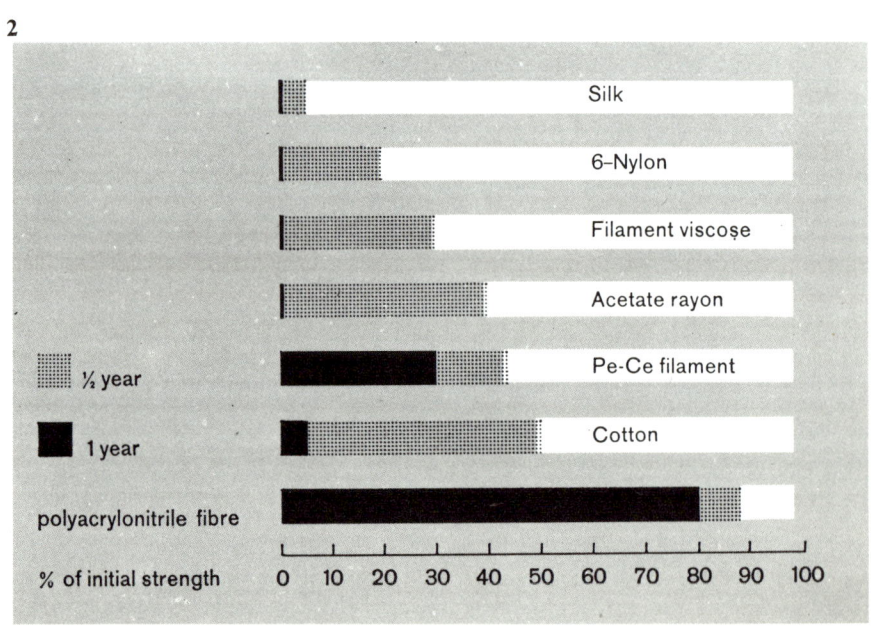

2

Divided rectangle

1 Diagram from a business publication, showing sales of various chocolate products for two different years. The size of the rectangle shows the increase in sales. Division of the rectangle is on a percentage basis and the parts are clearly labelled. But is 15% of the 1938 sales more or less than 8% of the 1962 sales? *Industrial Challenge*, Cadbury Brothers

2 Diagram from a scientific house-magazine, showing weathering resistance of various fibres. *Ciba Review*

3 Diagram from a paperback, showing world urban population by size of urban centre in the mid-1950s. The width of each horizontal rectangle is proportional to the population of the area it represents. This diagram highlights the difficulty of comparison of areas in this kind of diagram. For instance, in towns with over one million, does Germany have a larger total than Southern Asia? Good labelling can partly overcome this problem. J. P. Cole, *Geography and World Affairs*, Penguin Books

4 Diagram from a paperback, showing characteristics of earth's land area. Two aspects of the same problem are shown, neatly related by the use of the third dimension (see page 140). Anthony Barnett, *The Human Species*, Penguin Books

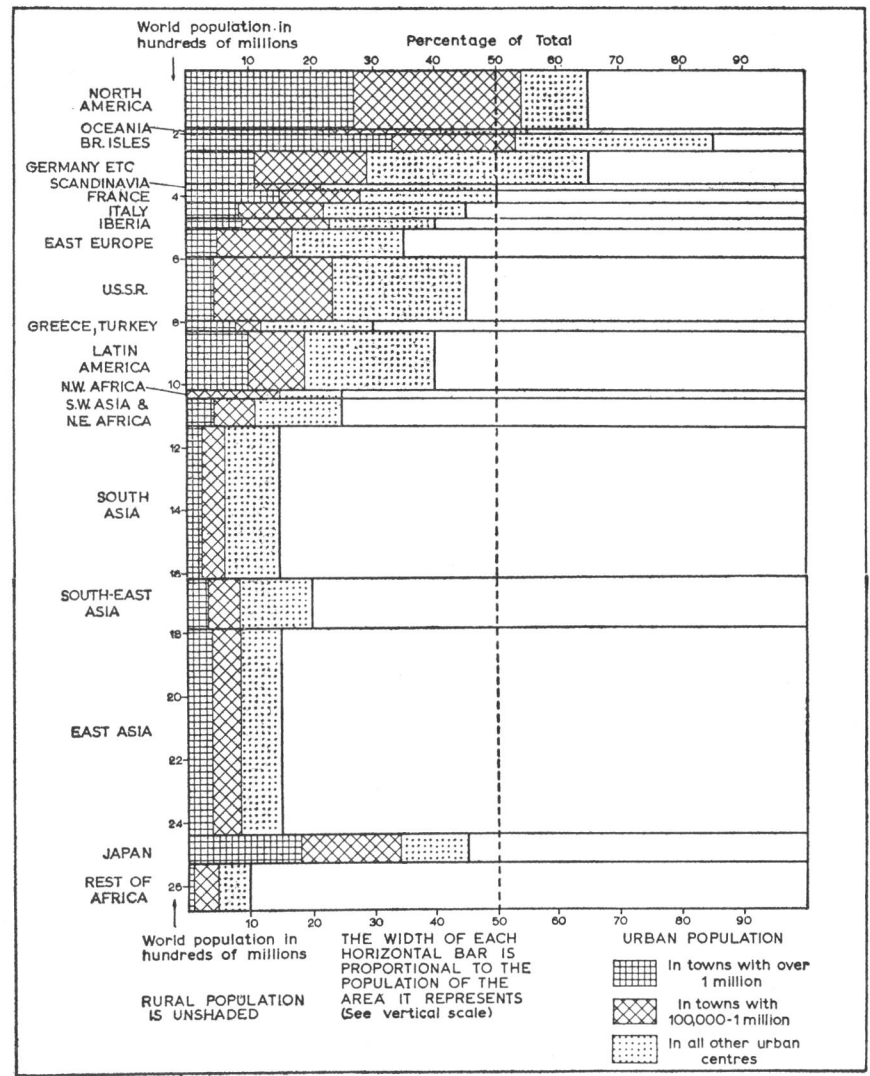

3

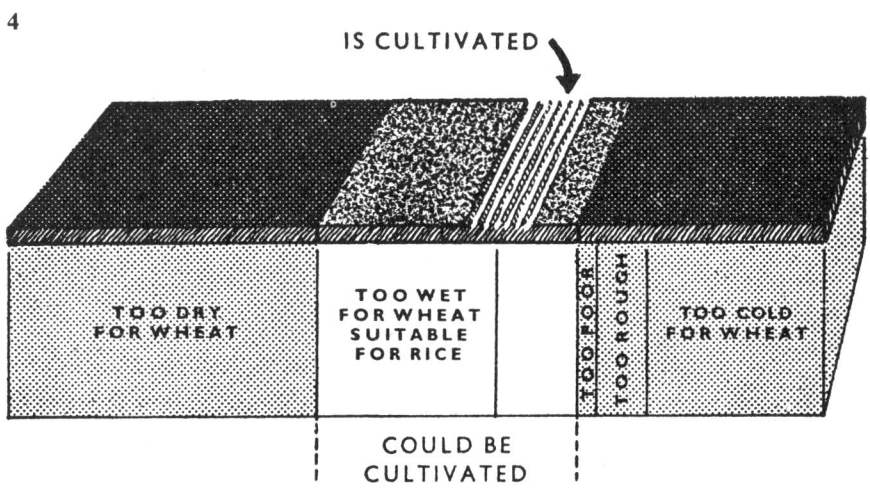

4

Circular graph

Figures which recur regularly, e.g. rainfall or temperature figures, can be plotted on a circular graph. Where a weekly or yearly cycle takes place, the normal line or bar graph may have disadvantages, as the left and right hand edges break the continuity of information. When this information is shown in a circular form, the problem is overcome. Values increase as they radiate outwards, and zero is represented for ease of drawing by a small circle.

The disadvantages of a circular graph are the problem of quick understanding by the reader who normally reads along a straight line or at the most through 180°. This graph requires reading and relating information through 360°.

There are variants on the circular form of graph, the circular bar graph and the wind rose being the most common.

Circular graph paper is available to help plotting problems.

1 Graph from a text book, showing temperatures month by month for Budapest. F. J. Monkhouse and H. R. Wilkinson, *Maps and Diagrams*, Methuen

2 Diagram from a text book, showing employment of working hours on a weekly basis for a group of sample farms in Finland. Known as an Ergograph, it shows the amount of work done at various times of the year. Ergograph is a specialist term coined by A. Geddes. F. J. Monkhouse and H. R. Wilkinson, *Maps and Diagrams*, Methuen

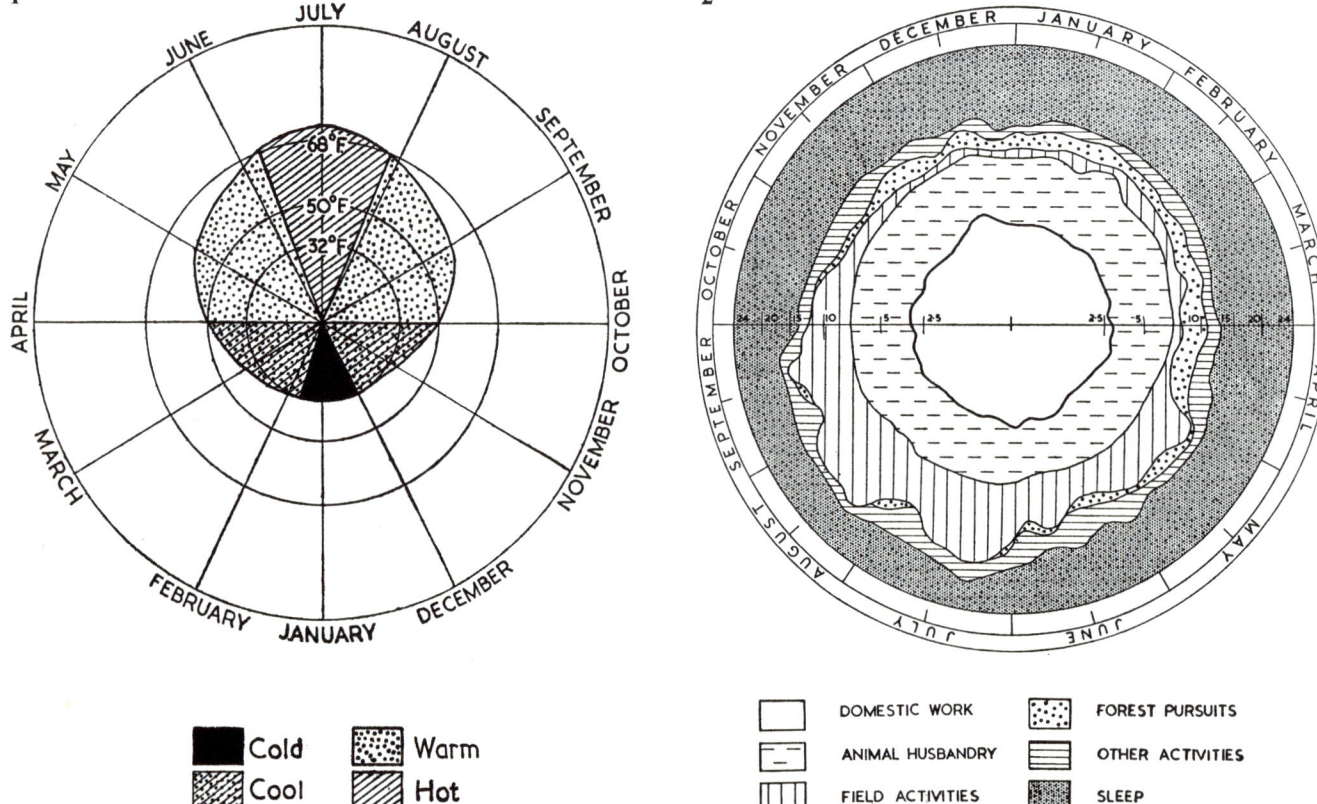

3 Graph from a text book, showing mean monthly temperature and rainfall at Fez, Morocco. Line and bar combined on circular graph. G. C. Dickinson, *Statistical Mapping and the Presentation of Statistics,* Edward Arnold

4 Diagram from a newspaper, showing maximum and minimum airport-to-city journey times for twelve European centres. *Observer*

5 A wind rose presents in visual form wind data for a location over a given period. The lengths of the arms indicate or what proportion of the period (say, a year) wind blew from each direction. The divisions of the arms analyse the winds according to velocity. There are various kinds of wind roses. Essentially a specialist diagram for the geographer or geographical map. E. G. Sterland, *Energy into Power,* Aldus Books

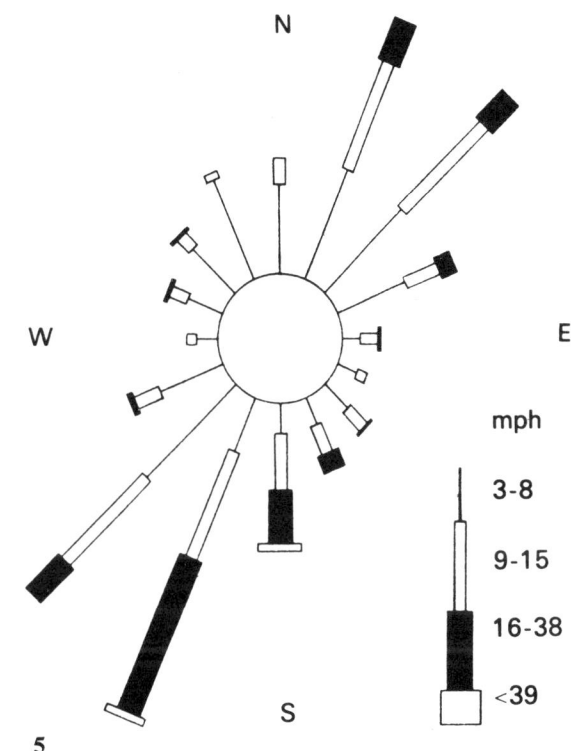

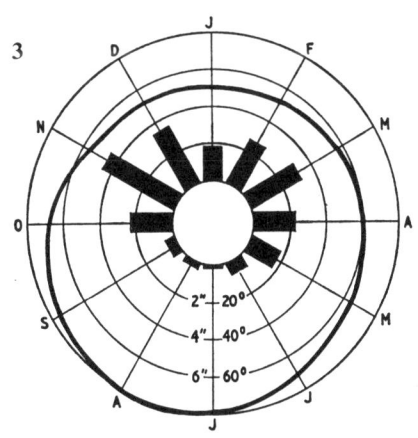

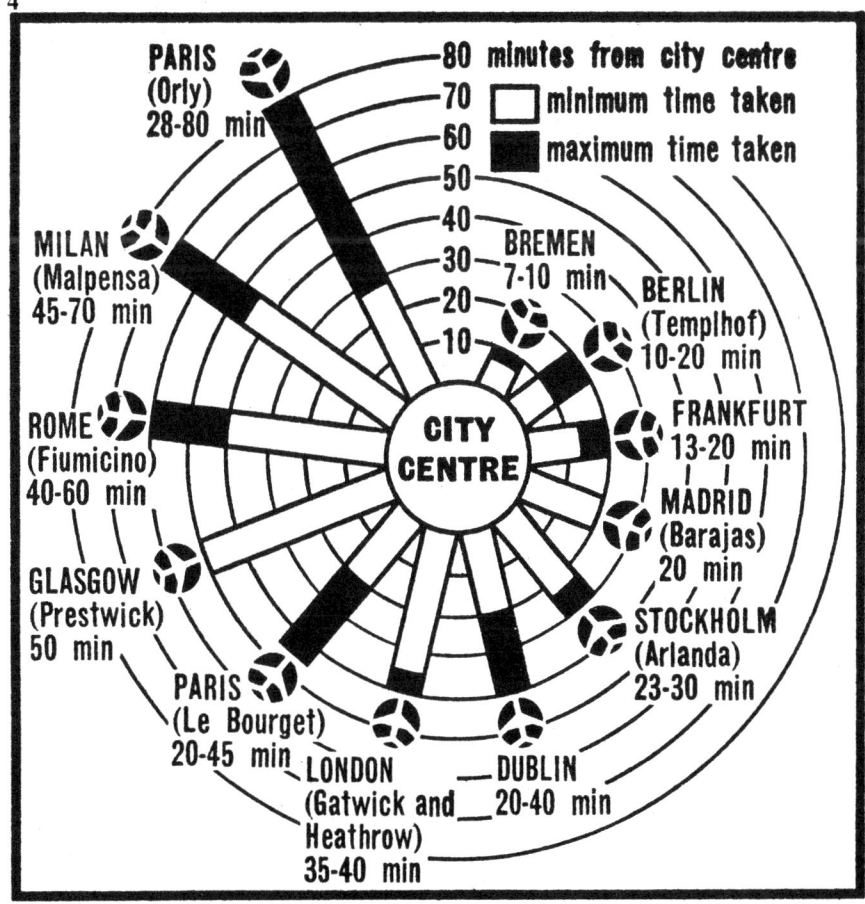

Divided circle (pie or cake graph)

The circle is divided into sections as in cutting up a pie. Hence the alternative name of pie or cake graph. Each section is proportional to the value it represents. If only two or three divisions are required the division of 360° can be calculated. For more complicated data it is more convenient to work with percentage circular graph paper.

The diagram has disadvantages both on the plotting as well as on the reading side. It is complicated to divide parts of a circle in proportion. There is always the possibility of error in calculation or plotting, even when using tables and special graph paper.

For the reader it is more difficult to compare small variations in the sections, and if the circles themselves vary in size, comparison of areas can be impossible.

Labelling is a major problem with the divided circle and ingenious if often misleading solutions are evolved to avoid turning the page round in order to read the names.

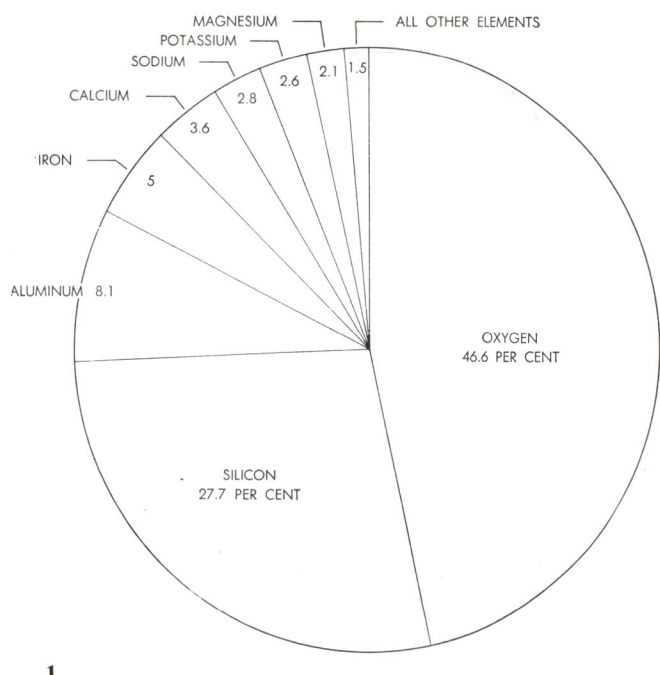

1

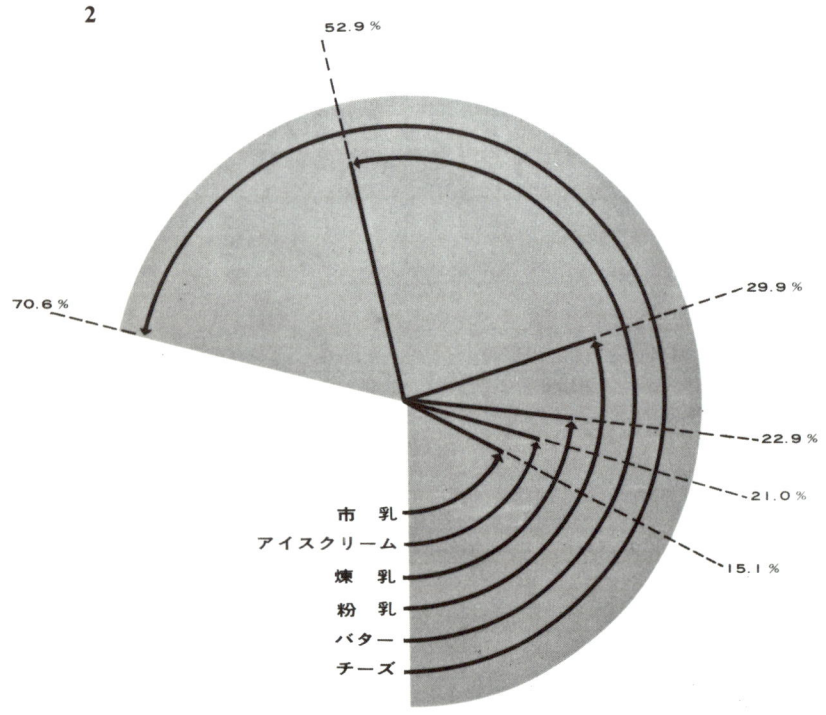

2

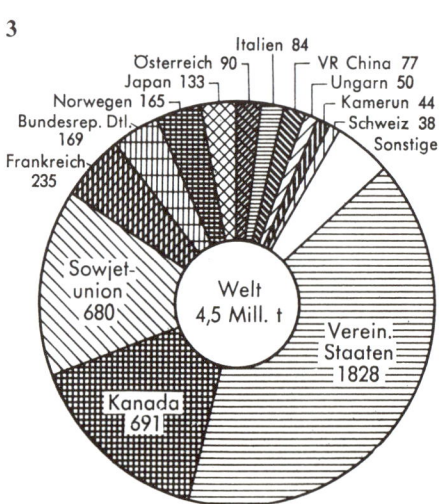

3

Aluminiumerzeugung 1960 (in 1000 t)

1 Diagram from a scientific magazine, showing composition of the earth's crust in weight. *Scientific American*

2 Diagram from a Japanese publication. Another solution to the labelling problem.

3 Diagram from a paperback encyclopedia, showing production of aluminium. Labelling and texture efficiently done. *dtv-Lexikon,* Deutscher Taschenbuch Verlag

4 Diagram from a magazine, showing the divided circle being used to record the time taken to get a meal at various restaurants on motorways. The divided circle is perfect for recording figures in minutes or hours provided they do not exceed twelve hours. *Drive,* Automobile Association

5 Part of a diagram from a book, showing increasing leisure. Original in colour. The circle represents 24 hours of a day! Divisions show work, leisure and sleep. Overlapping obviates repeating exactly identical circles. Norman Crosby, *Full Enjoyment,* Nicholson and Watson

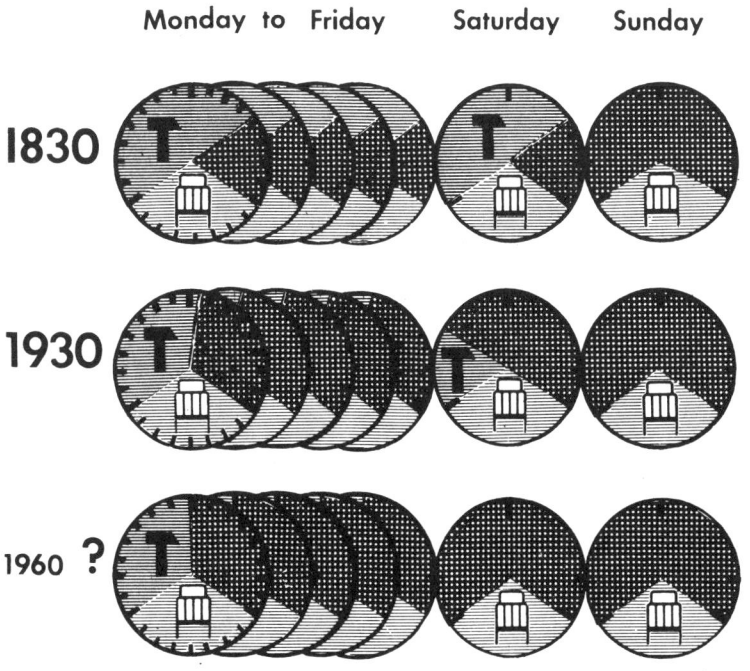

4

5

Increasing Leisure

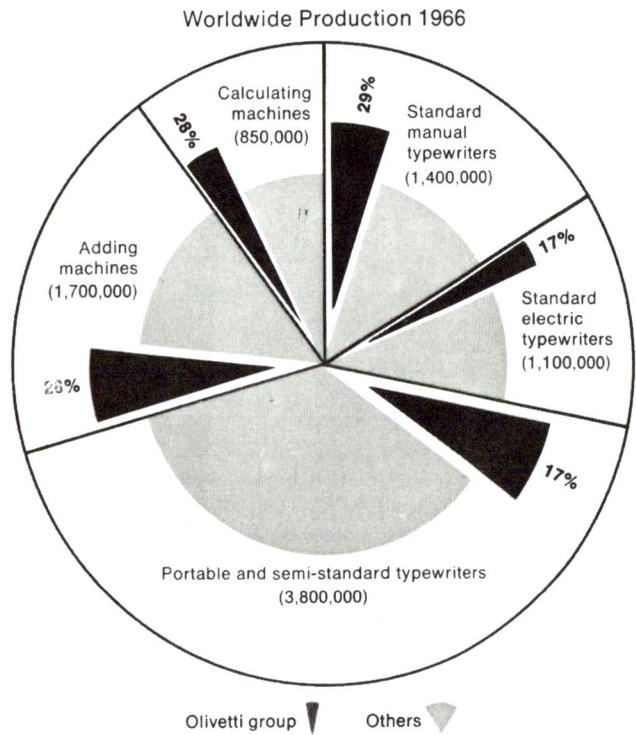

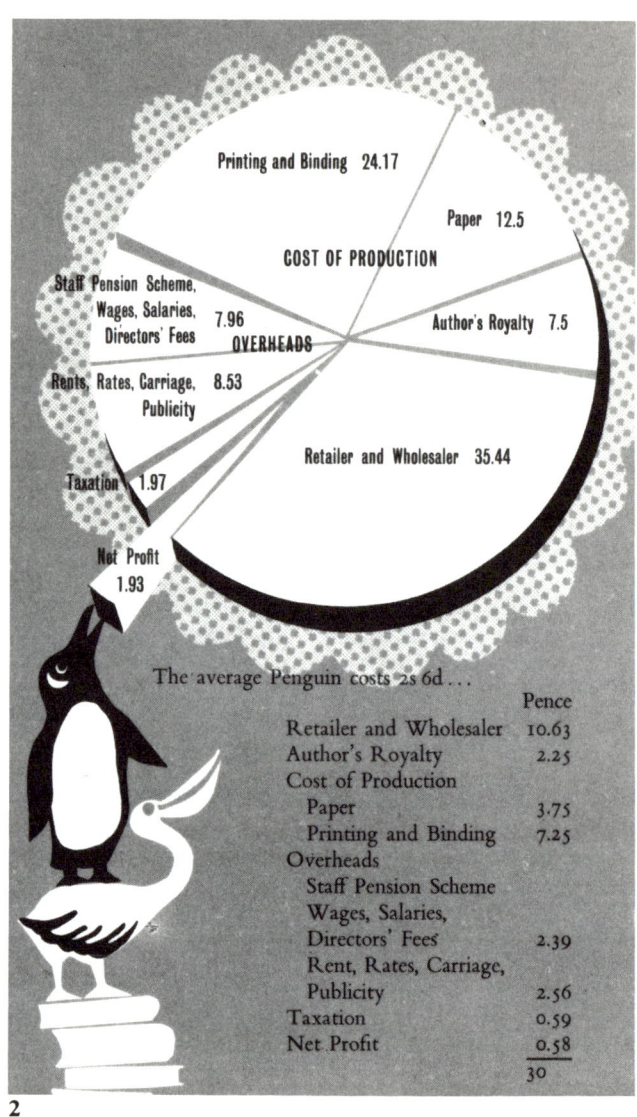

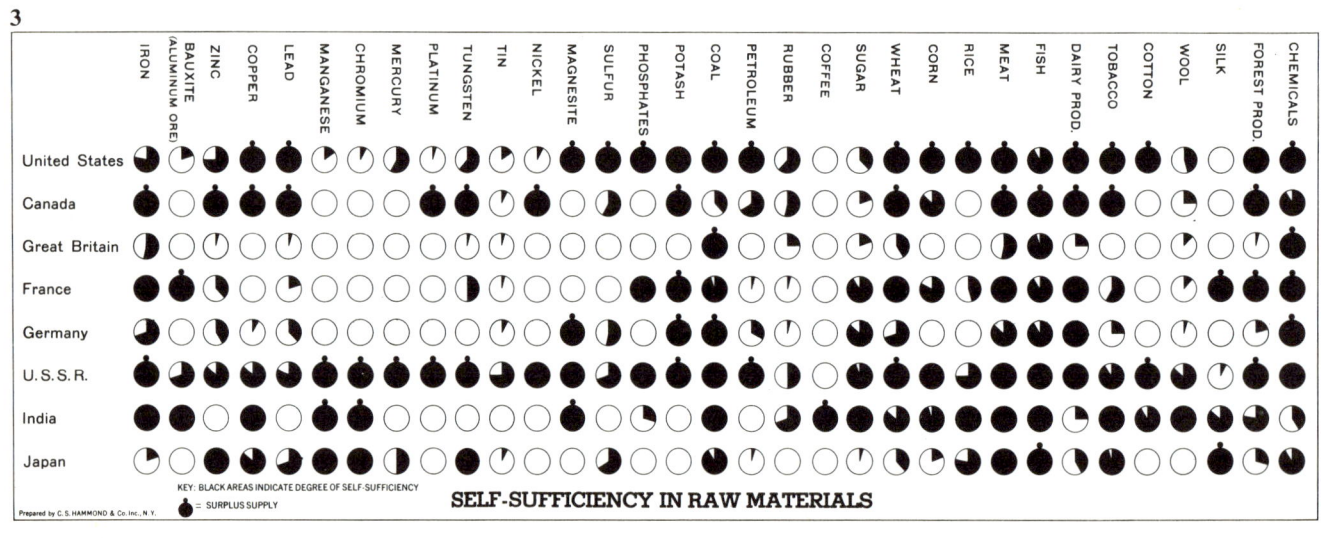

Divided circle

1 Diagram from a business magazine, showing world-wide production of office machines and the Olivetti share of each group. *Fortune*

2 Diagram from a book, showing the average cost of a Penguin book in 1956. The cake is cut and the small slice of profit shown. *The Penguin Story*, Penguin Books

3 Diagram from a school atlas, showing sufficiency in raw materials. Lack of material or surplus supply is shown. This is the divided circle being used as part of a tabular diagram (see page 128). *March of Civilisation*, Hammond

4 Diagram from a magazine, showing expenditure by governments on the First World War. *History of the 20th Century*, Purnell

5 Diagram from a magazine, showing population explosion. The discs represent world population at fifty-year intervals from 1850. The top disc shows the estimated figure for AD2000. The original is colour coded. Excellent use of three dimensional divided circles, where impact has more importance than information. *History of the 20th Century*, Purnell

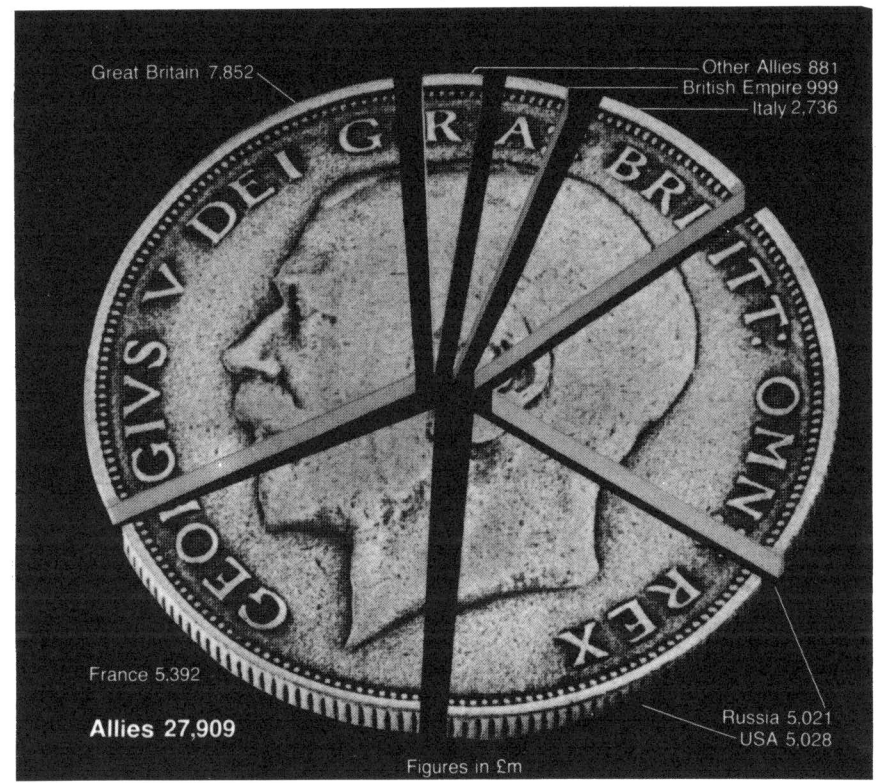
4

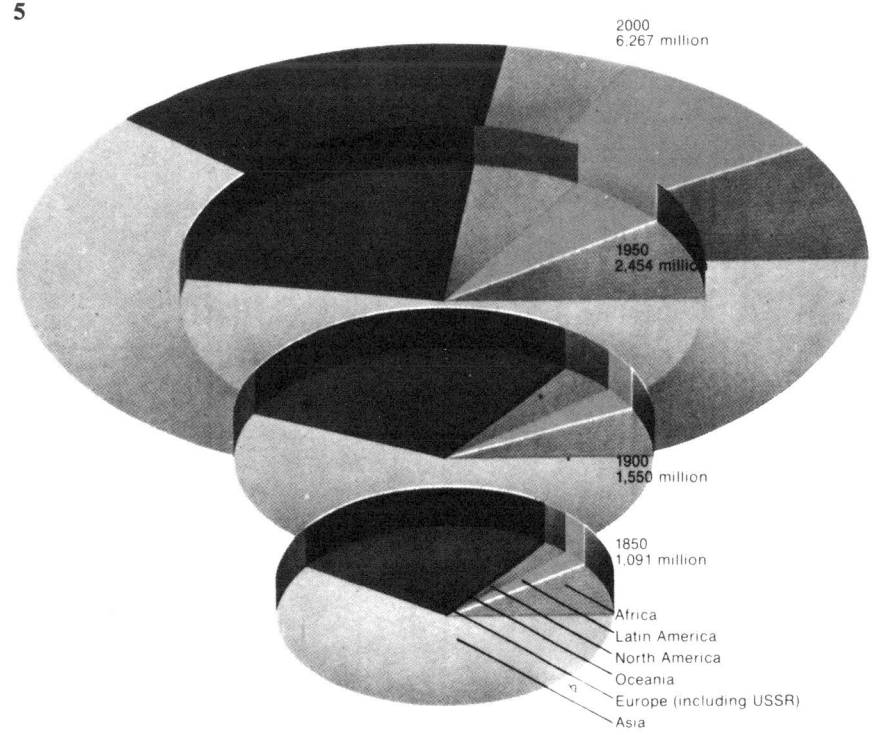
5

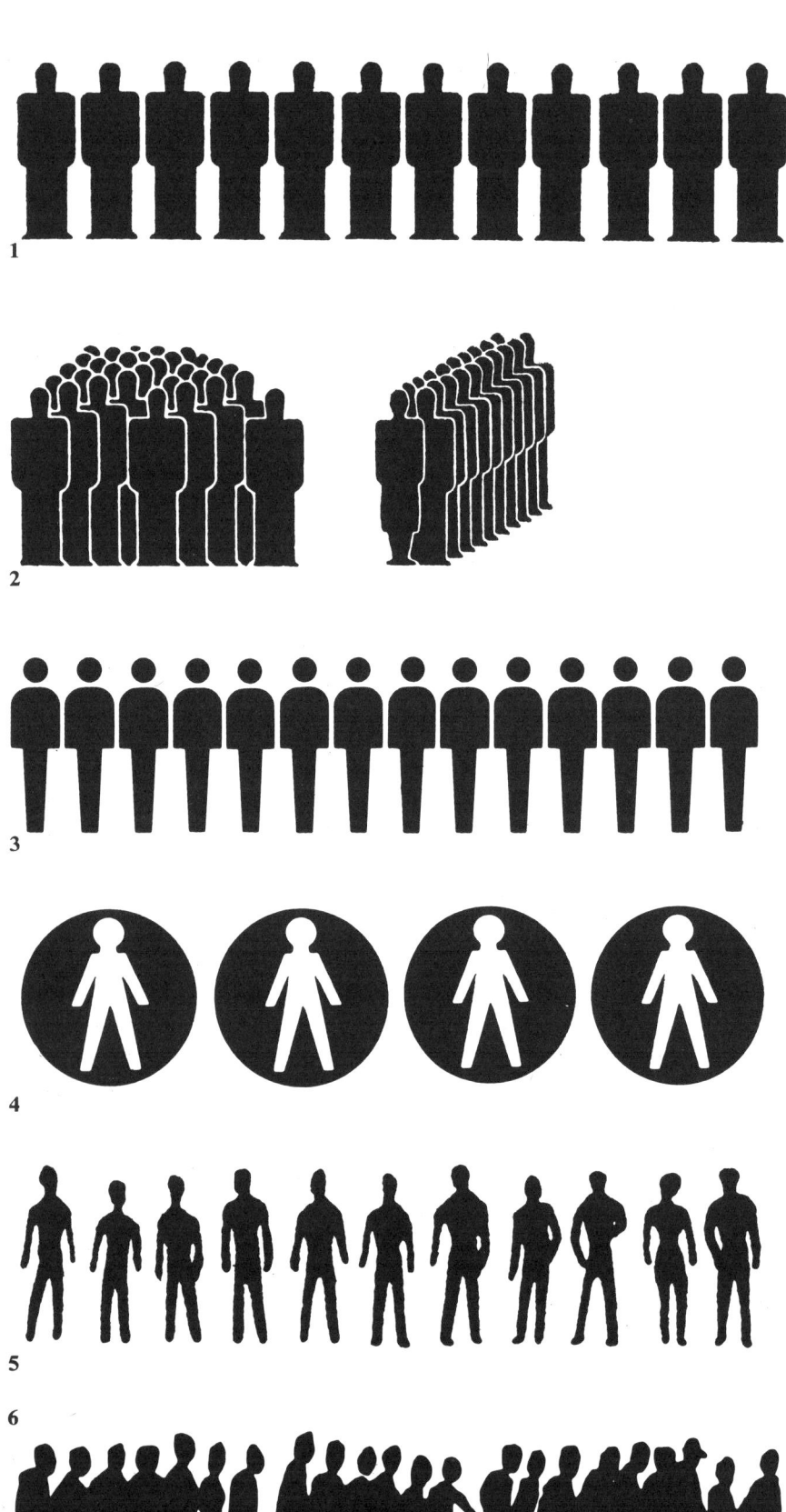

Since the important advantage of pictorial graphs is to 'humanize' diagrams the symbol for man is of prime importance. Here are a few examples:

1 Isotype man

2 Isotype 'crowds'. Because of problems of space, large numbers of people may need to be represented and this can be done in several ways.

3 Japanese

4 English

5 Italian. Freedom of approach – this is closer to a drawing than to a symbol, but it is still being used to represent a number of people.

6 English. This is no longer a symbol representing a number of people but an illustration of people queuing, and it is the length of line that represents the number (see diagram page 30).

7 Isotype from a school text book, showing world population. Marie Neurath, *Living with One Another,* Parrish

8 Isotype from school text book showing home and factory weaving in England. The interrelation of three distinct facts is clearly shown: how much cloth was produced, how many people worked at home, how many in the factory. The pictorial graph used to advantage. Think how this would look as a line or bar graph. Marie Neurath, *Living with One Another,* Parrish

Pictorial graph

The bar graph is non-representational. It shows very clearly the statistical information about a specific subject, but the actual subject usually has to be read in the heading or is concealed in the caption. The pictorial graph is an adaption of the bar graph, and is used to give the same kind of information. But it aims to overcome the lack of appeal of the bar chart by actually representing the subject.

The pioneer of the application of pictorial techniques to presenting statistics was Otto Neurath. He sought a way of conveying statistical information so that it would be readily understood by all people, irrespective of language and cultural barriers. He achieved this by developing simplified figures such as a man, woman, child, cow etc, and using them to represent information. He called these figures 'Isotypes' (I-nternational S-ystem O-f TY-pographic P-icture E-ducation). Variations of original figures are now part of the communication technique used by designers throughout the world. An Isotype is a picture diagram or symbol. The picture is used to represent a certain number of people or objects. The disadvantages of this representation come when fractions are involved, e.g.

if represents 100 men,

how are 225 shown? and 275 shown?

If rounding off numbers is accepted, and the approximation done consistently, Isotypes can be used with advantage. Ideally, the symbols must all be the same size and spaced equally so that measurement can be done quickly and accurately.

One important advantage of the pictorial graph is that, where numerous different kinds of items are being compared, it is clear from the symbols what items are being listed. A disadvantage may creep in when the symbol is so obscure as to confuse the reader.

Any increase in amount is best shown by repeating the symbols. When the information is simple there is a temptation to use the alternative means – increasing the size of the symbol. But this may produce problems of communication, and it is possible to interpret such a method as meaning that size rather than quantity has increased. Another temptation is to use one symbol only, and to shade/colour/cut proportionately to the amount it represents. This again can have the most misleading results.

World Population

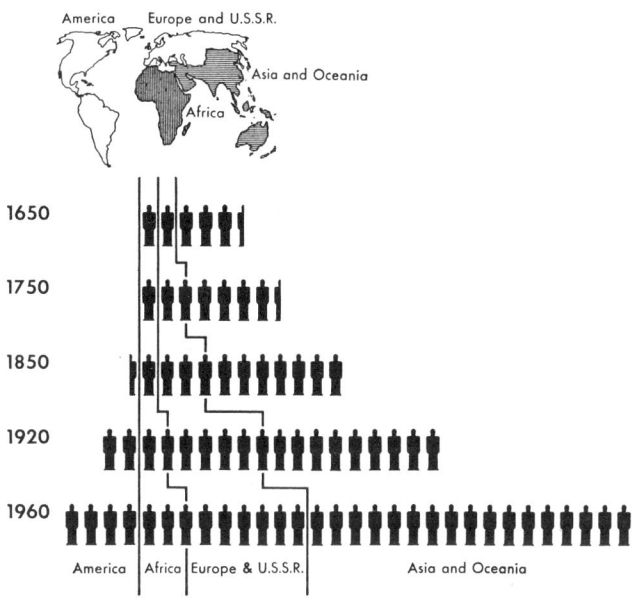

Each figure represents 100 million population

7

8

Home and Factory Weaving in England

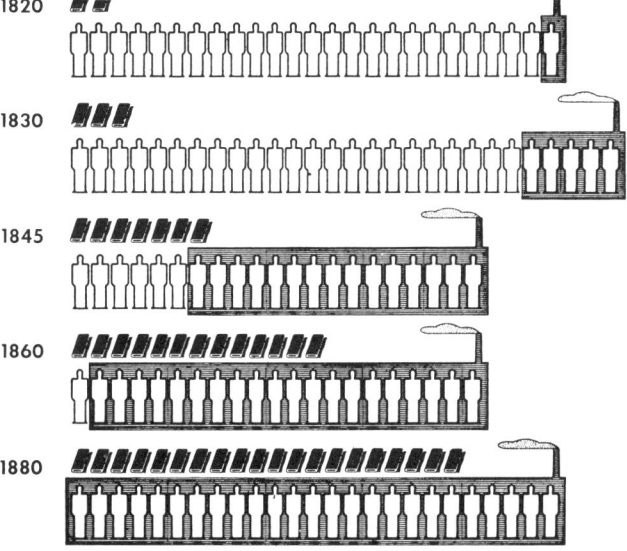

Each black symbol represents 50 million lbs. total production
Each man symbol represents 10,000 weavers
 outside factory : home weavers
 inside factory : factory weavers

Ownership of Wealth in Great Britain

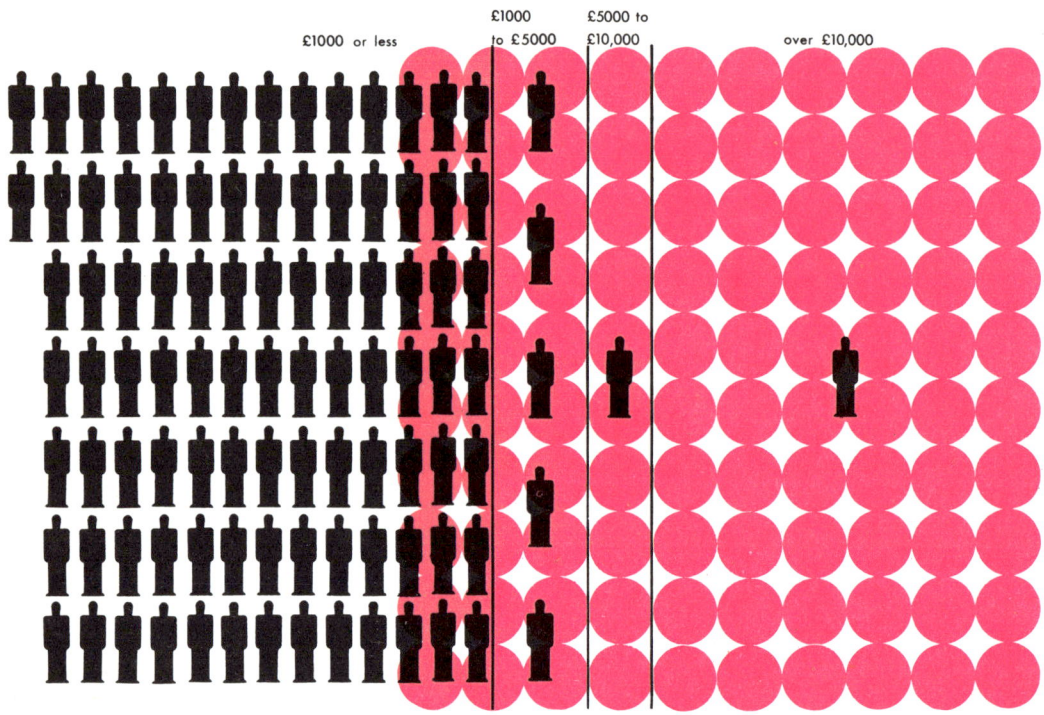

Each red circle represents 1% of total capital held
Each man symbol represents 1% of total population aged 25 and over

1

2
Distribution of Annual Incomes in Great Britain

Each group of symbols represents 1 million income receivers about 1941 (single symbols 100,000)

black: incomes under £250
blue: average income £300 and £500
green: average income £1,000 and more

Pictorial graph

1 Isotype from a book, showing wealth in pre-war Britain. Man (representing 1% of the population) overlaps circle (representing 1% of capital held). Michael Young and Theodor Prager, *There's work for all*, Nicholson and Watson

2 Isotype from a book, showing annual income in Britain in 1941. The Isotype crowd in use. Both diagrams, from the same book, show dramatically the way two elements were related, the first by area, the second by height of column. Michael Young and Theodor Prager, *There's work for all*, Nicholson and Watson

3 Isotype from a book, showing world population grouped according to area and race. The symbol for man has been modified and colour-coded to represent races. L. Secor Florence, *Only an Ocean Between*, Harrap

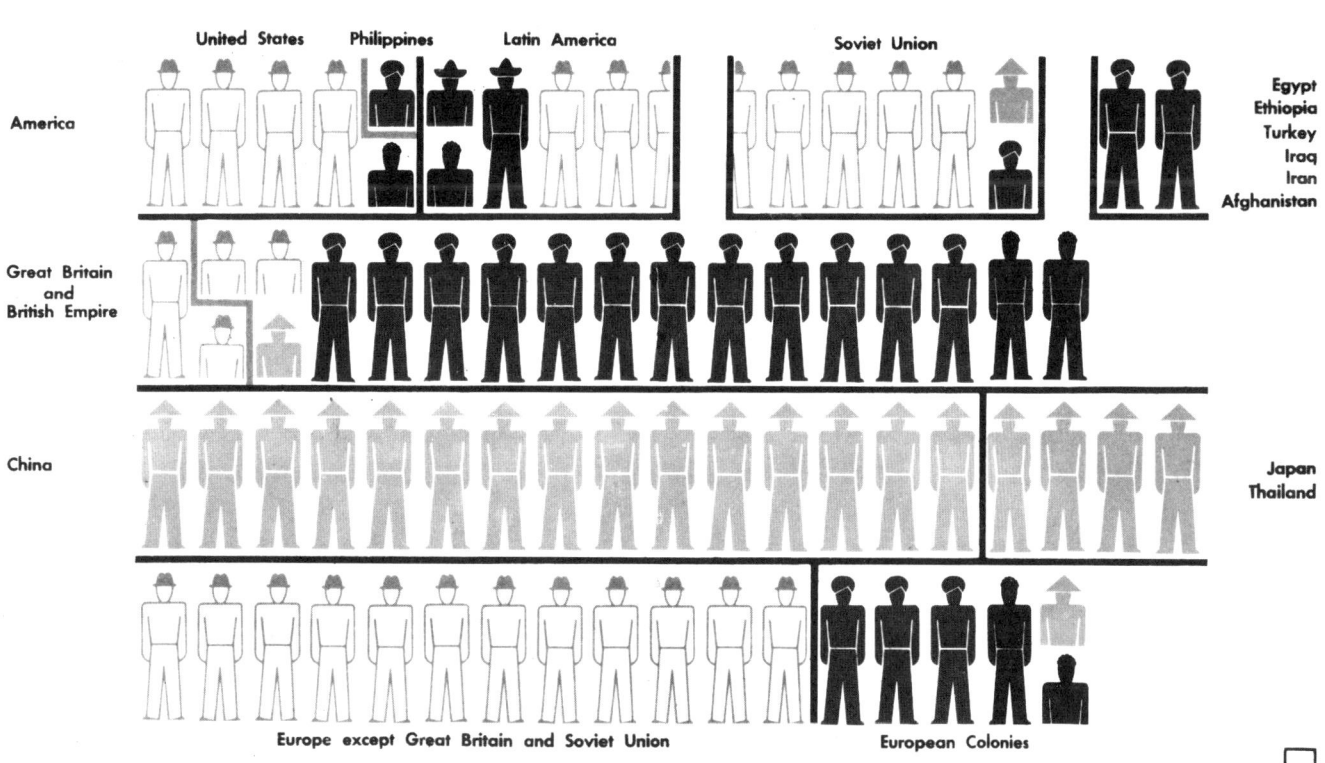

59

Pictorial graph

1 Diagram from a school atlas, showing travel time between Pittsburgh and Philadelphia. The clock represents four hours of travel. *American History Atlas,* Hammond

2 Diagram from a school atlas, showing shift of population from rural to urban areas. The symbol for man has a different dress according to place and period. *American History Atlas,* Hammond

3 Diagram from a school book, showing home-produced and imported food for UK. Placing symbols in a box partially overcomes the problem of rounding off numbers to nearest half symbol. Paul Redmayne, *Britain's Food,* Murray

4 Diagram from an encyclopedia, showing 'who-produces-what'. Each symbol represents one-tenth of the total world production of each major raw material. It shows for example that America produces 40% of the world's cotton. By locating the symbols geographically, the diagram clearly shows the trade that has to go on between various areas. *Man in Society,* Macdonald

Each clock represents 4 hours of travel

2   1

100 90 80 70 60 50 40 30 20 10 0
HOME PRODUCED
IMPORTED
FATS
0 10 20 30 40 50 60 70 80 90 100

3

4

North America
Europe
USSR
Latin America
Africa
Asia
Oceania
Middle East
cotton
raw wool
flax fibre
raw silk
rubber
synthetic
s
n
natural
coal
crude oil
roundwood
iron

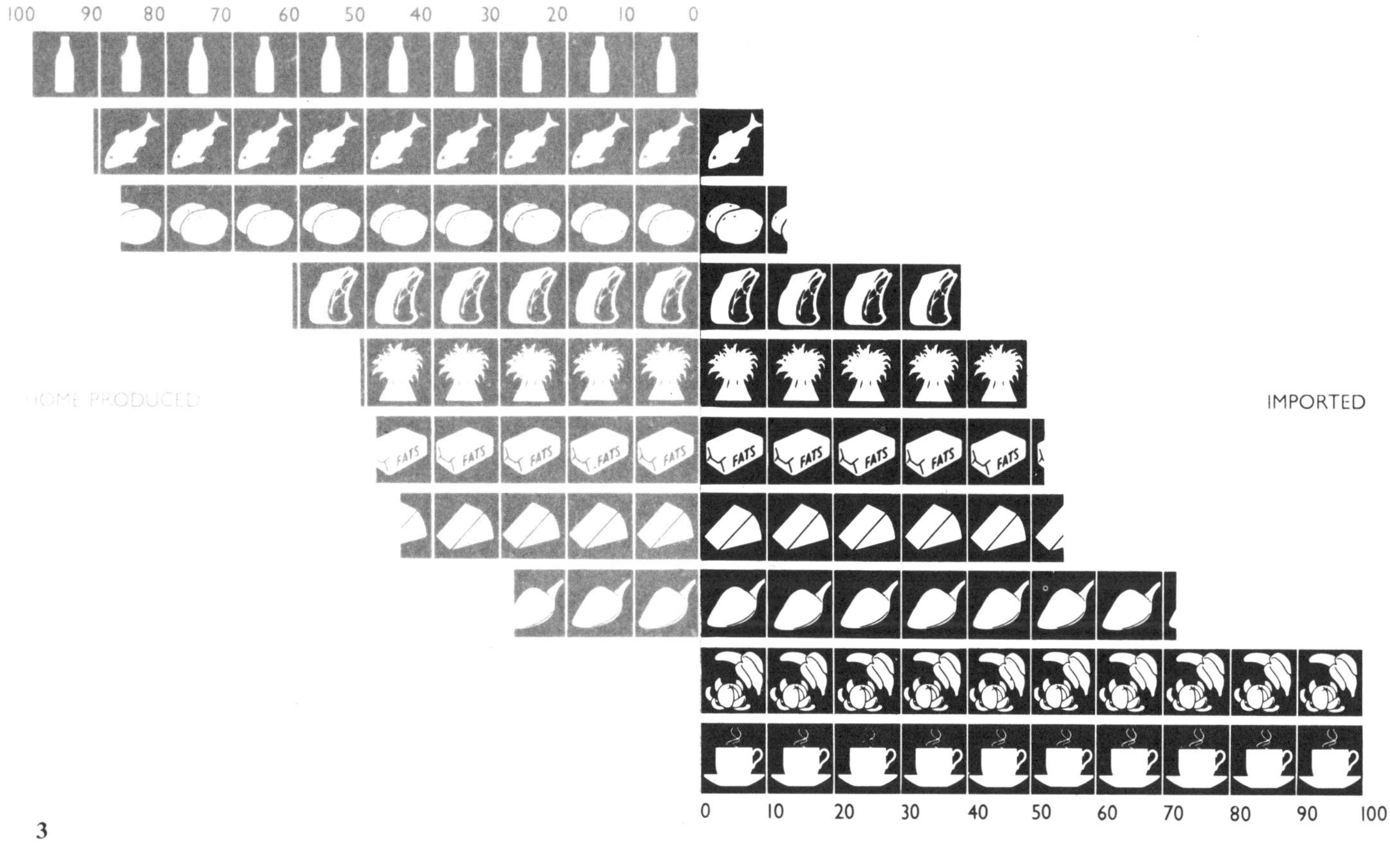

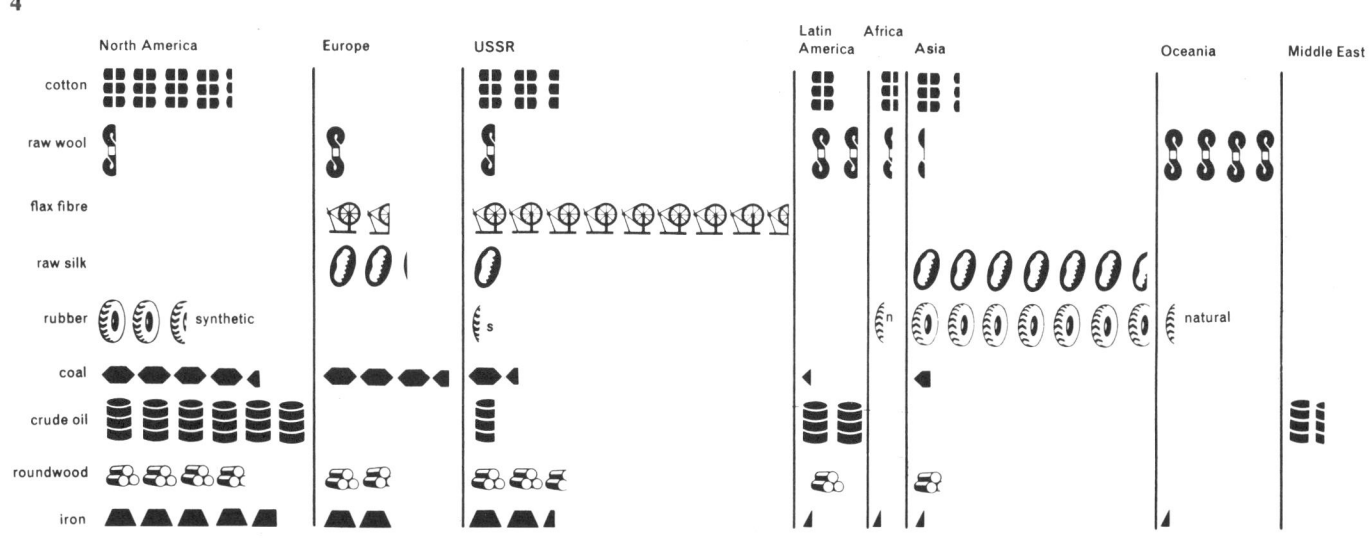

TYPES OF DWELLINGS IN GT. BRITAIN   RENT OF A COUNCIL HOUSE

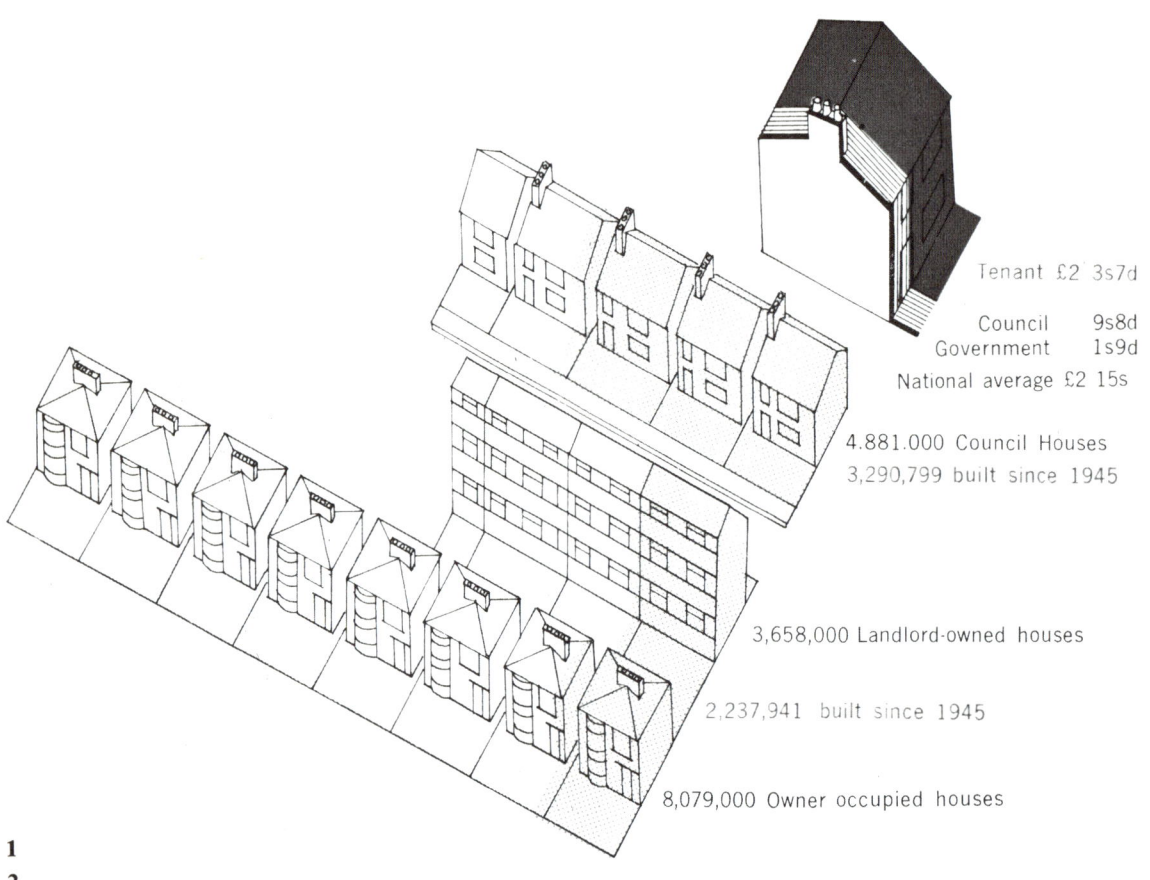

Tenant £2 3s7d
Council 9s8d
Government 1s9d
National average £2 15s

4,881,000 Council Houses
3,290,799 built since 1945

3,658,000 Landlord-owned houses

2,237,941 built since 1945

8,079,000 Owner occupied houses

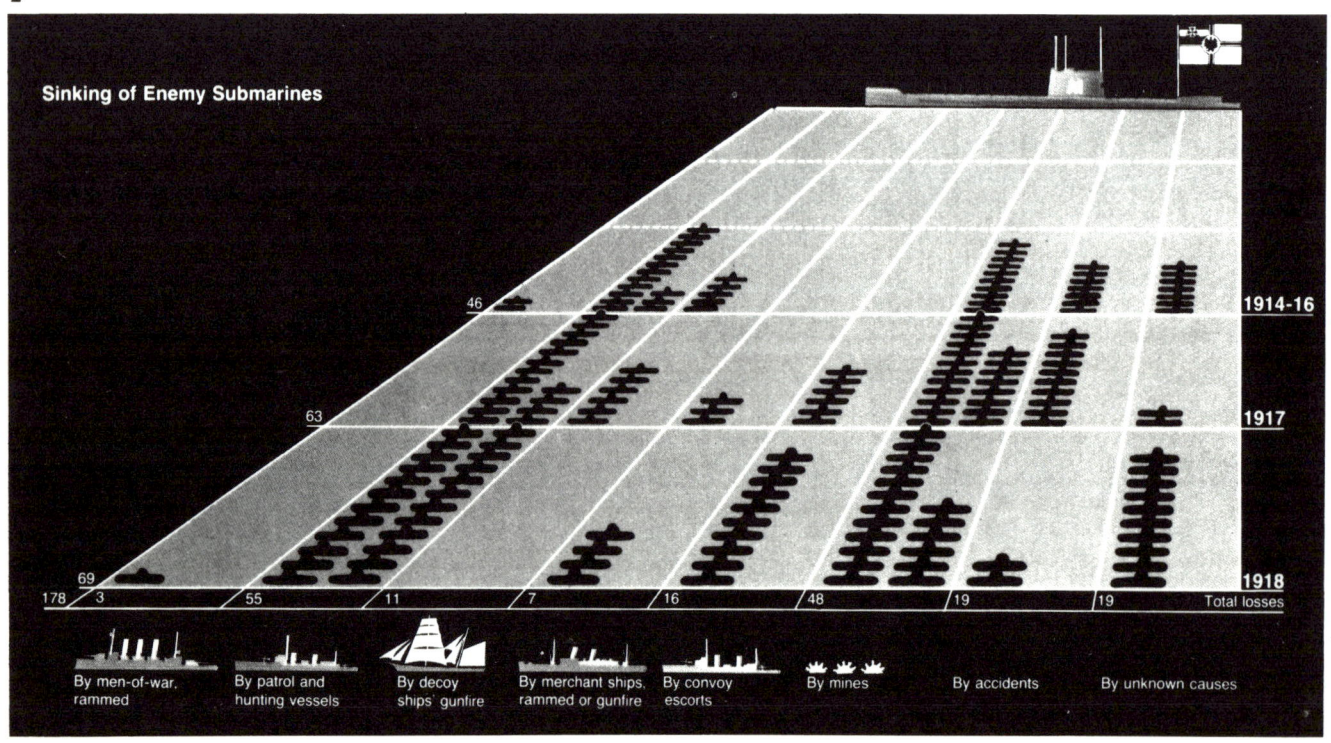

Pictorial graph

1 Diagram from a school publication, showing three types of dwellings in Britain (area shaded shows those built since 1945), and the rent of a council house. BBC Publications

2 Diagram from a magazine, showing how enemy submarines were sunk in the First World War. *History of the 20th Century,* Purnell

3 Advertisements showing the number of flights which Air Canada makes to Canada each week. The pictorial graph is favoured for publicity since the information is easily transmitted without lengthy labelling or captions. The 'representational' approach appeals more than the abstract one of the bar diagram (page 40) or tabulation chart (page 128). Air Canada

4 Diagram from a magazine, showing the maximum speed achieved by man and animals over a short distance. How realistic should a pictorial graph or diagram be? *Drive,* Automobile Association

3

4
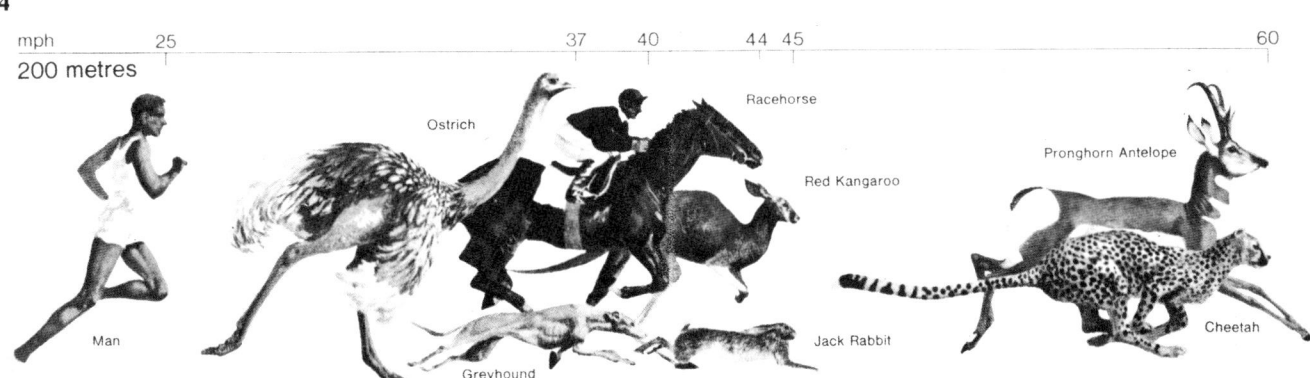

Pictorial graph

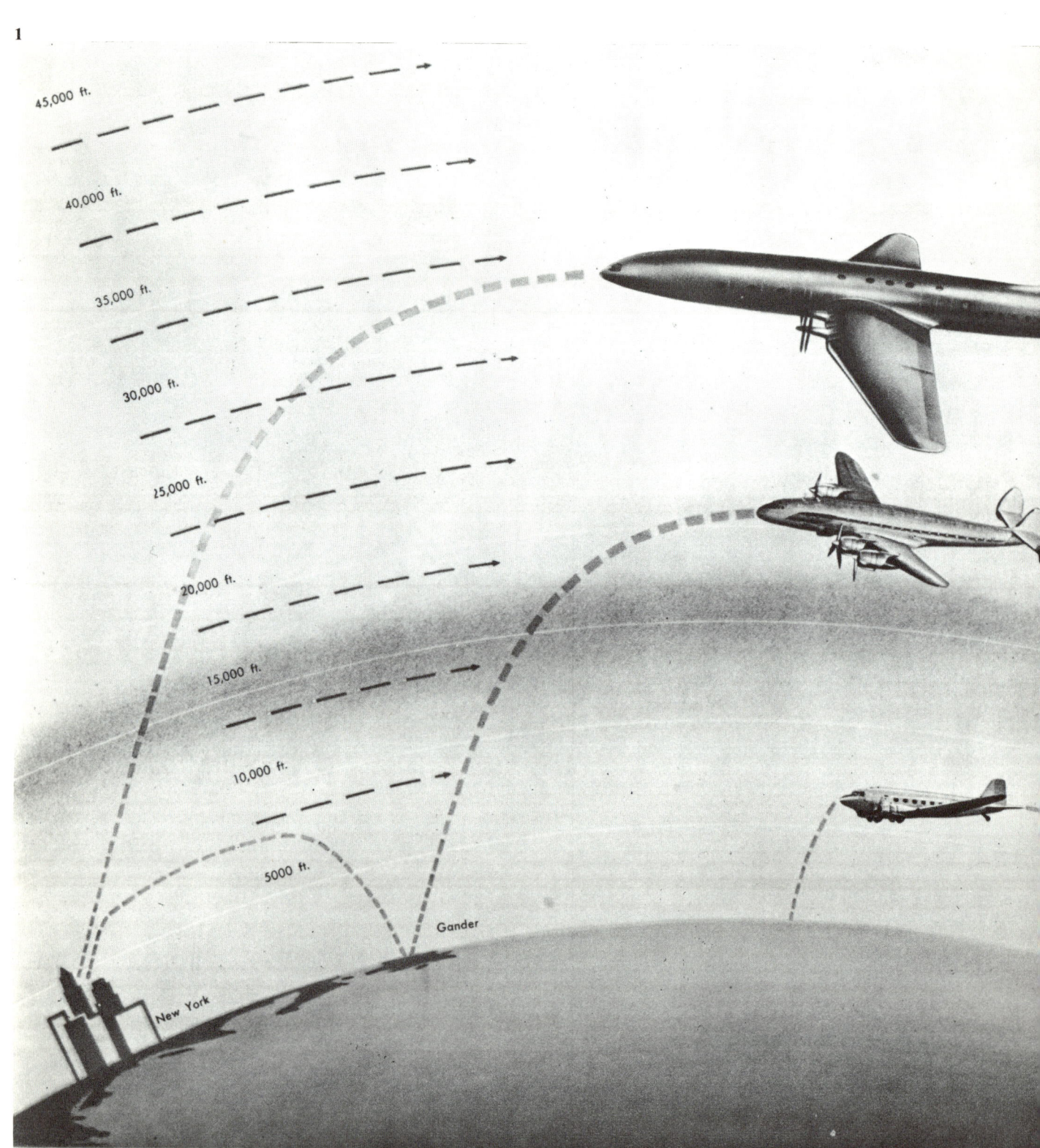

64 Pictorial graph

1 Isotype from a magazine, showing the differences of three aircraft. This shows the ability of the Isotype approach to cope with a variety of information and give it visual coherence and enable the essential message to be understood, quickly. The dashed line shows the range of the three aircraft. The arrows on the left represent the head winds faced by the planes at various heights. The symbols on the right represent horse-power, weight (of engine, fuel, structure, payload) and people carried (crew, passengers). Colour coding sorts out the groups more clearly. *Future*

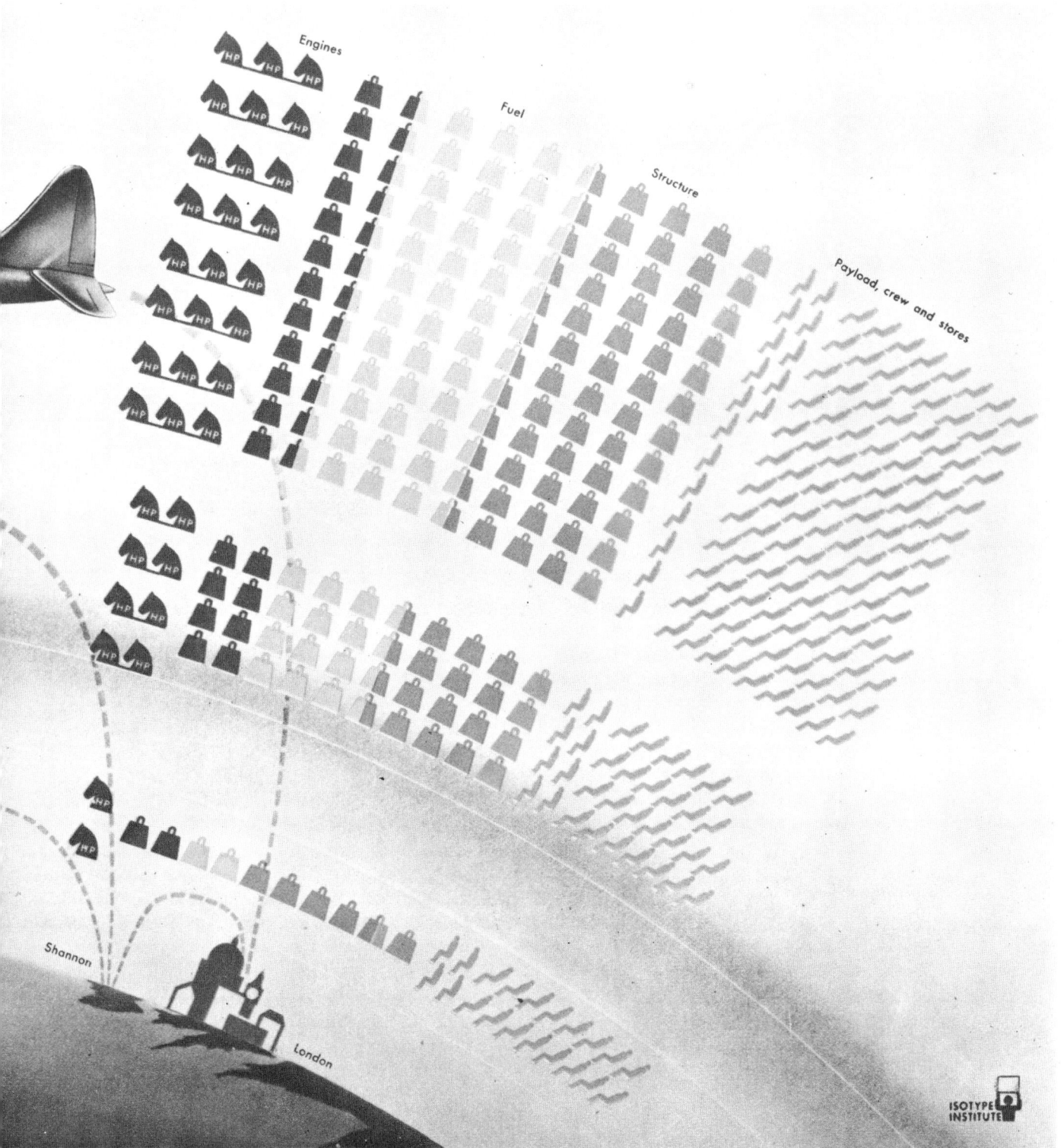

2 Explanatory and statistical maps

This part of the book deals with maps, explanatory and statistical. In discussing maps, the many problems which face the cartographer can only be hinted at. The main concern is the kind of situation which the graphic designer might have to sort out. In the bibliography on page 143 I recommend books dealing with the wider aspects of cartography.

Again it must be said that the illustrations come from a variety of sources with different aims in mind, so that various solutions are shown for the methods described. Not all maps therefore come from atlases – they are from newspapers, text books, motoring publications and other sources. This influences considerably the way in which facts are selected, how the design approach is organized, and the amount of information included on the map.

The most important difference between the statistical diagram and the explanatory/statistical map is that the map provides a basic framework on which information can be shown. In giving statistics in diagrammatic form there may be several approaches open to the designer and whatever diagram method is used, there will be a certain freedom to decide the actual shape and size.

But once a map form is introduced there is a 'basic grid' which has considerable influence on the final appearance of the diagram. One very obvious problem is getting information of a similar nature into areas with considerable variation in size and shape. This can be further complicated when labelling, for both map and diagram, is required. To superimpose population pyramids for various countries on a map of Europe may be easy for France and Spain but impossible for Luxembourg and Holland if the map lines are to be kept readable.

Map projections

There is no ideal map projection. A projection which is good for central America may not be good for northern Asia. A projection used for a navigation chart may not be suitable for a statistical map.

On *equal area* maps, each part, as well as the whole, has the same area as a globe at the same scale. But one result is distortion of shapes and angles.

On *conformal* maps, any small area has the same shape as on a globe and any one point is correct directionally, with regard to any other point which is near to it. On a conformal map of the world it is the scale which is distorted. The Mercator projection is conformal – compare the size of Greenland with South America. Conformal maps are suitable for navigation, as directions are not greatly distorted.

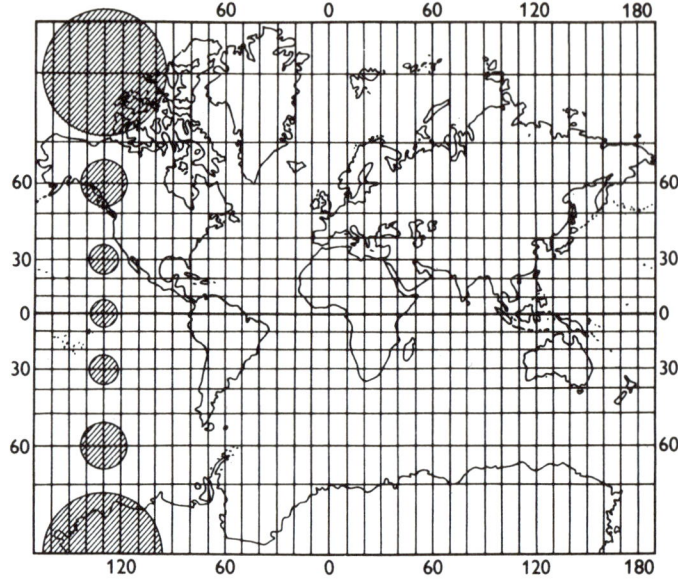

The Mercator projection has very good shapes for small regions, but the scale is enormously exaggerated towards the poles. It is advisable not to use the map beyond latitudes 75° N. or S. In spite of this, Mercator is widely used for world maps. It fills a page well and the exaggeration of northern regions can be positively beneficial if, for example, a country such as Norway needs to contain much information or labelling.

Gall's projection, similar to Mercator's, exaggerates the northern land less, and is used in many British atlases.

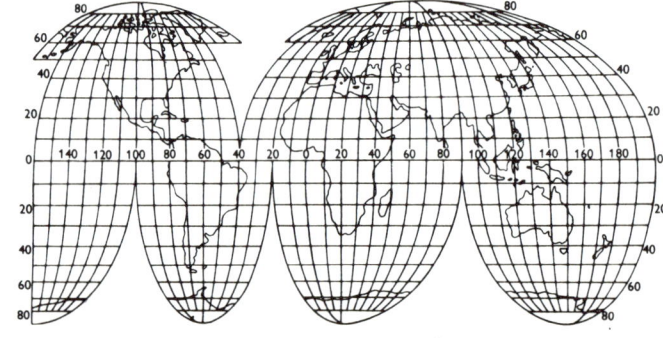

Goode's interrupted projection is an equal area map. It combines good shape with equality of area, but since the oceans are separated it is less good at showing global relationships. It could not be used, obviously, for international shipping or air routes.

There are other projections commonly used, and still others which have been devised to show a specific view of the world, such as the projection centred on the North Pole.

There are three important considerations which affect the designer when choosing a projection for a map:

1 the purpose of the map;
2 the position on the globe of the map area to be drawn – tropical, polar etc;
3 the extent of the map area – large or small, or whether it appears as a long strip going north to south or one going east to west.

In maps showing distribution, such as population density, equal area is the important factor. In maps for navigational purposes the first consideration is the preservation of direction.

Explanatory maps

There are two kinds of maps dealt with here, explanatory and statistical. In many cases they are combined on one map and certainly some techniques discussed apply to both.

Route maps

The most common type of explanatory map is the route map, ranging from a map showing routes of explorers to a more diagrammatic form showing the best road to a particular place.

Historical route maps are concerned with accurate representation of land and sea areas, so that the route distances may be drawn in correct proportion to each other.

Several routes may be shown on the one map. This is the same problem as the graph with several lines – to devise a good system of line coding so that each route may be recognized and can show clearly the intricacies of the journey, especially when crossing other routes.

Maps for present-day use vary considerably. The normal road map has good representation with simplification, and its presentation of information is based on the aim of showing the best road system. An extreme form is the road map as a diagram of the motorway, consisting of a straight line with the sequence of exits, service areas etc, marked on it (page 72).

Ray map

The ray map shows lines radiating out from a central point to various other points. These lines can show spheres of influence or merely record distances to surrounding points. Several centres and their lines of influence and the way they overlap can be shown on one map. The statistical version of this type of map is the flow line map.

Non-quantitative

There are other kinds of explanatory maps: political (page 78), physical (pages 82–3), distribution of natural vegetation (page 81), race, religions (page 95) and the weather map (page 69).

Maps can also give information about people, products and other things without any quantitative statement. It gives a general statement which summarizes a situation. A simple example of a non-quantitative map shows:

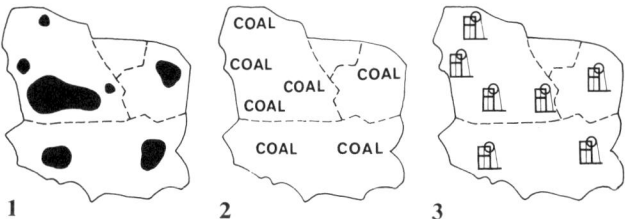

1   2   3

1 shading, giving extent of coal fields;
2 coal fields located by lettering;
3 coal fields located by symbols.

To select information from detailed sources, to translate or simplify for a new map, requires skill, if distortion is not to creep in. There is no statistical diagram or labelling to qualify the information on the map. What to put in or leave out is one of the problems of this kind of generalized map. Should the three small coal fields be left out to create a better representation of the relative size and importance of the coal fields? A statistical map showing production would not have the same problem.

Statistical maps

These maps show and locate statistical information. The map compares quantities of coal in an area or at particular places, or how much coal is sent to what destination.

These are three main ways of comparing such statistics on a map:

1 quantities located within an area;
2 quantities located at particular points;
3 quantities located along a series of lines.

Each method has its own techniques which shape and influence the appearance of the map.

The first way, that of quantities located within a given area, is the most common. Given an area, countries or country, statistical information can be shown in various ways. Summarized they are:

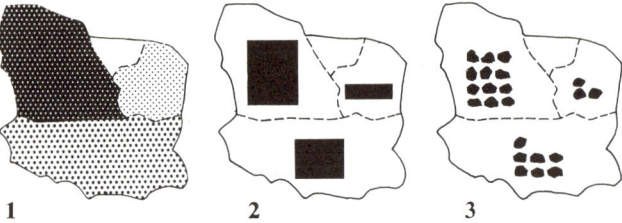

1 shading map area by colour, texture, proportional shading, repeated dots;
2 repeated diagrams within the area, e.g. bar graph, divided circle etc;
3 repeated symbols.

The second method, quantities located at a given point is achieved by using:

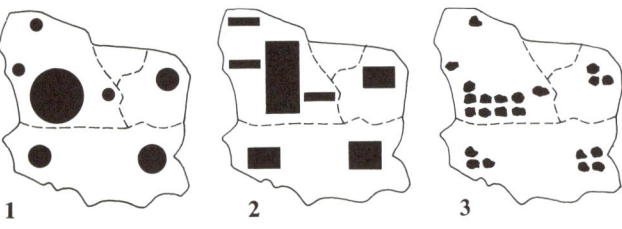

1 proportional circles, spheres, squares, bars, blocks;
2 repeated diagrams;
3 repeated symbols.

With quantity distributed along a line there is one method only in common use – that of a band proportional in width to the quantity, running from one point to another. This is the flow line map.

In reality many of these methods merge, and several may be used on one map. I have therefore grouped the illustrations according to appearance to make comparison of the methods easier. They may be explanatory or statistical, the statistics may be located within an area or at a given point. In this way one can see how the problems of each vary, and how methods may influence each kind of map. The captions bring out the relevant facts and the problems of the various techniques are now dealt with in more detail.

Techniques

Colour, line, tone, texture

These techniques for dealing with map areas apply both to the explanatory and statistical maps. Colour can be used both, for example, to show different political areas, or to show amounts of rainfall. But the requirements may influence the approach. In a map showing political divisions, distinct areas need to be shown, and colours are used to create this effect. But with a quantitative map, where there is only one subject being shown, it is usually advisable to limit the colour range, and get variation by using several tones of each colour, rather than a 'rainbow' selection of colours. It is important to understand the psychological properties of colour, red being notably dominant, blue being cool and recessive, grey being neutral. Where several colours are used, it is usual to go from one to another in the order of the spectrum. Where quantities are large, the colour chosen must dominate by tone and colour, while small quantities should be recessive or neutral. One qualification to this must be made concerning the size of areas to be shown – areas of small quantities are tiny, and colour which is bright but light in tone will probably be the answer.

Line is limited in its use to explanatory statements about the physical characteristics of area. Various cartographical systems have been devised to show relief, ranging from simply drawing contour lines to hachure (page 82). Relief can also be shown by flat tone or graduated tone, representing mountains and valleys.

Tone can be used for both explanatory and statistical statements. As a method by itself it is limited, since the maximum number of clearly differential tones is probably four or five. If they are in a continuous gradation, more tones might be used, but if they are separated it is not possible to detect differences except where the variation is considerable.

Line textures are the most widely used method of shading map areas. The range of texture is enormous, and some of the considerations involved in using colour are true for texture. Large numbers need to be a dark dominant texture, small numbers to be of a light, quiet texture. Textures can be used which merge into one another or which are completely different. Textures can be suggestive of what is being represented (see vegetation map page 81).

The main problem of line texture is labelling. Cut-out white areas for letters to fit into are not satisfactory. It is possible to produce a variety of textures by using broad bands, large squares, dots, all in a light tone. Lettering can be read on such a background.

Another problem with textures (and in fact with colour and tone) is how to show two areas merging, or having an indefinite boundary. With tone, the edge or edges can be faded; while colours can be intermixed by stipple or line shading. With line texture where this problem occurs, it is best to choose a texture which will suggest an indefinite edge (broad pattern of dots), or which will allow an overlap to show up clearly by contrast.

Weather map

Weather maps use internationally recognized symbols. The lines on these maps, called isobars, join points of equal atmospheric pressure. Triangles shown along a line represent a cold front, half circles represent a warm front. Most weather maps used in newspapers are kept simple and do not go beyond this kind of information. But there are symbols for cloud cover, wind strength, rain, snow, fog, mist, etc.

1 Weather map, designed for use on television, showing isobars, cold and warm front. British Broadcasting Corporation

2 Weather map designed for television giving more detailed information about Britain; used in conjunction with the previous map. Though symbols are based on the international system of weather symbols, not all are immediately comprehensible. A dot for rain is clear but what does the inverted triangle mean? British Broadcasting Corporation

3 Weather map as it would appear on a television screen.

Television imposes its own discipline on map (and diagram) design. Areas have to be shown by different shades of grey. Large areas of white must be avoided because they vibrate on the screen. Use of texture has to be carefully controlled and is best avoided to prevent a clash with the lines of the screen. Lettering should be large and simple. Fine lines on light backgrounds tend to disappear. Relief maps of the kind shown on page 75 can be used effectively. A map, appearing on the screen for only a short time, must be understood in that time. It cannot be studied at leisure as in an atlas. It must therefore be kept simple and contain only essential information.

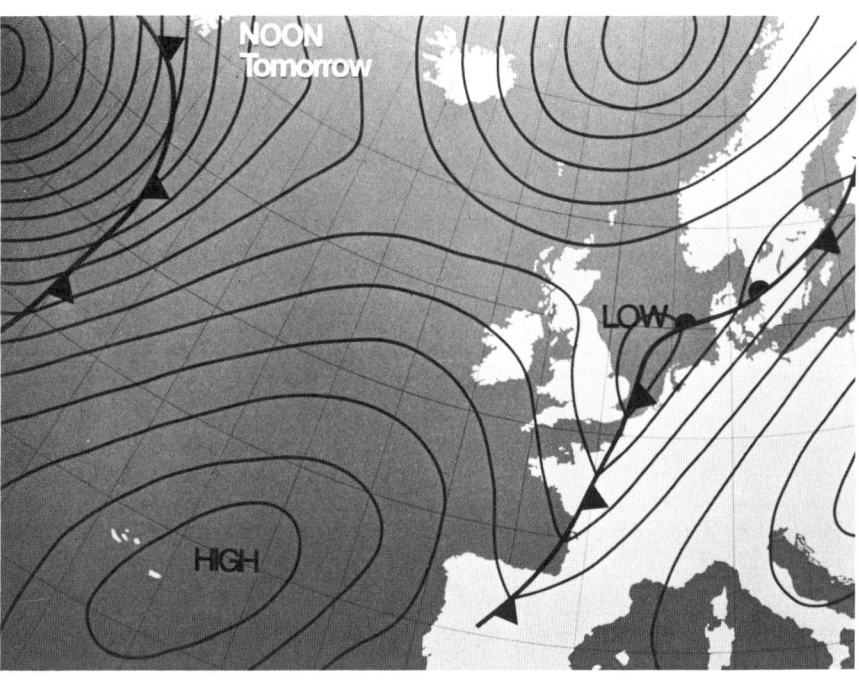

1

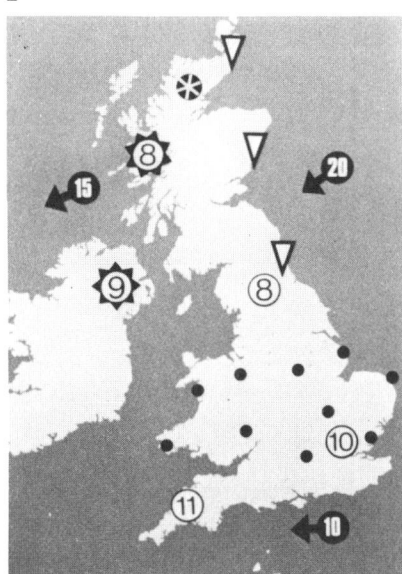

2

3

Bars, blocks and spheres

Bar graphs, separate bars of one graph, may be located on the area they represent. The map is a visual reminder of the location of each element. The problem is to locate correctly widely differing lengths of bar on a map so that they do not overlap, and especially to locate bars on a particular point. Since there is no common base line, comparison of length is difficult and good labelling is required.

If the length of bar proves impossible to fit in, proportional squares may be used. And if there is enormous variation of quantity, the advantages of the third dimension may be needed, and proportional cubes or blocks may be used.

Location of bars is normally used for quantities related to an area–i.e. where there is room to superimpose a bar or bar graph. Blocks and block piles are more useful for specific points, and the base can be clearly located on that point.

Dots, circles and spheres

Information to be represented by dots is placed on the area of the map concerned – simple idea but in practice full of problems which need to be resolved. One problem is the size of dot.

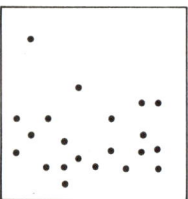  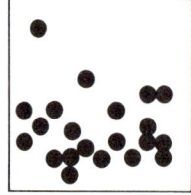

If the dot is too small, any variation of distribution is difficult to read. If the dot is too large, congestion is caused and there are large, uninformative areas. Dots should only just begin to join up in the densest areas.

Another problem is the placing of the dot in the area. This will depend partly on the informational requirements of the map, but there are two basic ways.

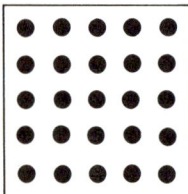 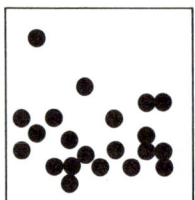 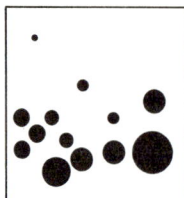

One is to spread the dots evenly over the area. If, however, the variation of distribution is known, an uneven placing of dots can be made. This leads one to ask whether one dot is sufficient to cope and whether a variety of sizes of circle would show the information better.

The dot map is widely used. It shows variations in distribution well, but actual quantities weakly. It is best used where marked variations are seen.

An alternative to the drawn dot is the round headed pin, and this attractive, dimensional method is used for magazines and display maps and diagrams (page 141).

Proportional circles are used on maps for quantities to be located at a specific point. The main difficulties arise from locating a large number of circles in a small area, and from recognizing different sizes of circles without measuring each one. A graduated row of circles in a key presents no problem (page 92), but the actual circles on the map lose their identity.

Spheres can also be used on maps in the same way as proportional circles. Their three-dimensional quality helps them to stand out on a 'flat' map.

Divided circles are used on maps, but the disadvantage is comparison of the size of sectors. If the divided circles vary in proportion and have complex divisions, the information is impossible to sort out. The simpler the divided circle the stronger it is shown on a map, and the greater the impact.

Symbols

As already suggested, symbols can be used in several ways; they can be used in an explanatory way, giving location or general information about an area, or, usually by repetition of symbols, they can make a qualitative statement.

The first problem is that of the design of the symbol so that it is instantly recognizable. Will a symbol locating something be the same as the symbol making a quantitative statement? For example, a shaft locates the coal mine. To show production does one repeat numerous mine shafts or use a symbol of a lump of coal?

Location of one symbol is not difficult but locating groups of symbols making a quantitative statement can present problems. Symbols for statistics about particular places are not easy, and symbols are mainly used for production figures, or import and export from countries, or larger areas.

Maps using symbols must have a key. The symbol should not require labelling underneath the symbol on the map, although in a few cases this might be done to avoid a long and involved key.

Flow lines

The map shows the exact route or diagrammatic direction followed by, for example, traffic, goods or money, with the width of line being used to convey quantitative information. The main problem is to get the width of line the right size so that they do not join when several are close together. The diagram is essentially linear and other information is not easy to add.

Other problems

Distorted map (cartogram)

The map is not always the absolutely fixed framework discussed earlier. It can be modified in various ways. A simple and familiar example is a road or rail map (such as that used for the London Underground) where the main representational element is the sequence of stations along each route. All maps are abstracted to some degree, but in the diagrammatic map or cartogram the distortion may have gone so far that the country may not be recognized.

In fact, it is probably only effective to distort certain maps which are very familiar, e.g. the World, the United States, Britain. Unless we know how the map normally looks, the impact of the way the information is being shown is completely lost. The most common use is distortion of area in proportion to population.

Latitude and longitude

In the same way that lines of graph paper are largely omitted from the line graph, so maps which present statistical information omit the complicated network of lines of latitude and longitude. But for some maps, such as ocean route maps, or maps where an unusual projection has been used, latitude and longitude lines are essential. It is possible to mark off the degrees on the frame of the map and avoid a network of lines on the map itself.

Title, key and scale

It was suggested with the line graph that the title of the graph should be prominently shown. This is as true of the map as of any other diagram which must stand by itself. But the tendency is to include the title as part of an informative caption.

The key is indispensable, for it provides the explanation to the symbols used. All symbols used on the map should be included in the key (possible exceptions are the obvious symbols, part of common cartographic practice: dots for towns, dotted lines for boundaries). Ideally the symbol should appear in the key, drawn exactly the same size and in the same manner as drawn on the map.

The key to statistical symbols, proportional circles, flow lines, is obviously essentially to understand and extract the information and should be shown prominently.

The scale of map varies in importance according to the use of the map. On maps showing routes or relationships that involve distance, the scale is an important factor in making the map useful. In certain cases, a scale of both miles and kilometres is an added advantage.

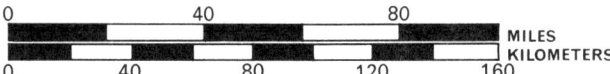

Route map

1 Map from a timetable, showing railway lines of the region concerned. Diagrammatic rendering, important lines shown thicker (this suggests a flow map – see page 104). *British Rail*

2 Map from a road atlas, showing a motorway. The key information for the motorist is the sequence of exits, service area, and the geography of individual exits. *The Reader's Digest AA Book of the Road*

3 Map from a guide book, showing the walk round Notre Dame. Route and building to be looked at are clearly shown. All other elements are left out. *Paris*, Michelin

4 Map from a road atlas, showing town centre of Leicester, England. Main road, one way streets, names and numbers as large as possible for the motorist to read quickly. Reproduced same size. *Members' Handbook*, Automobile Association

5 Diagram from a newspaper, showing the hospital where the wounded President Kennedy was taken. *Sunday Times*

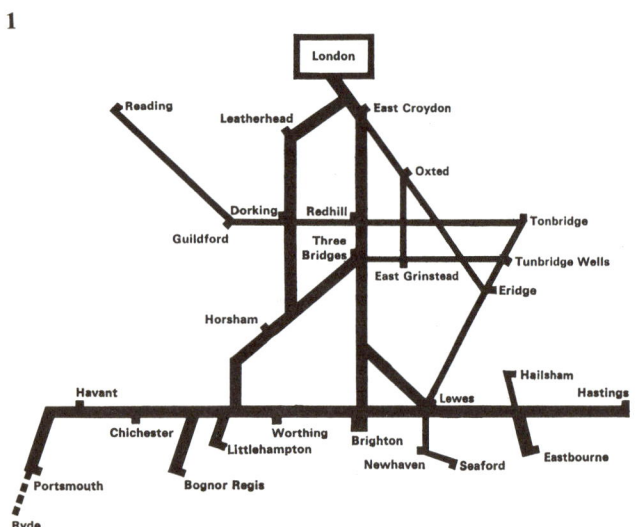

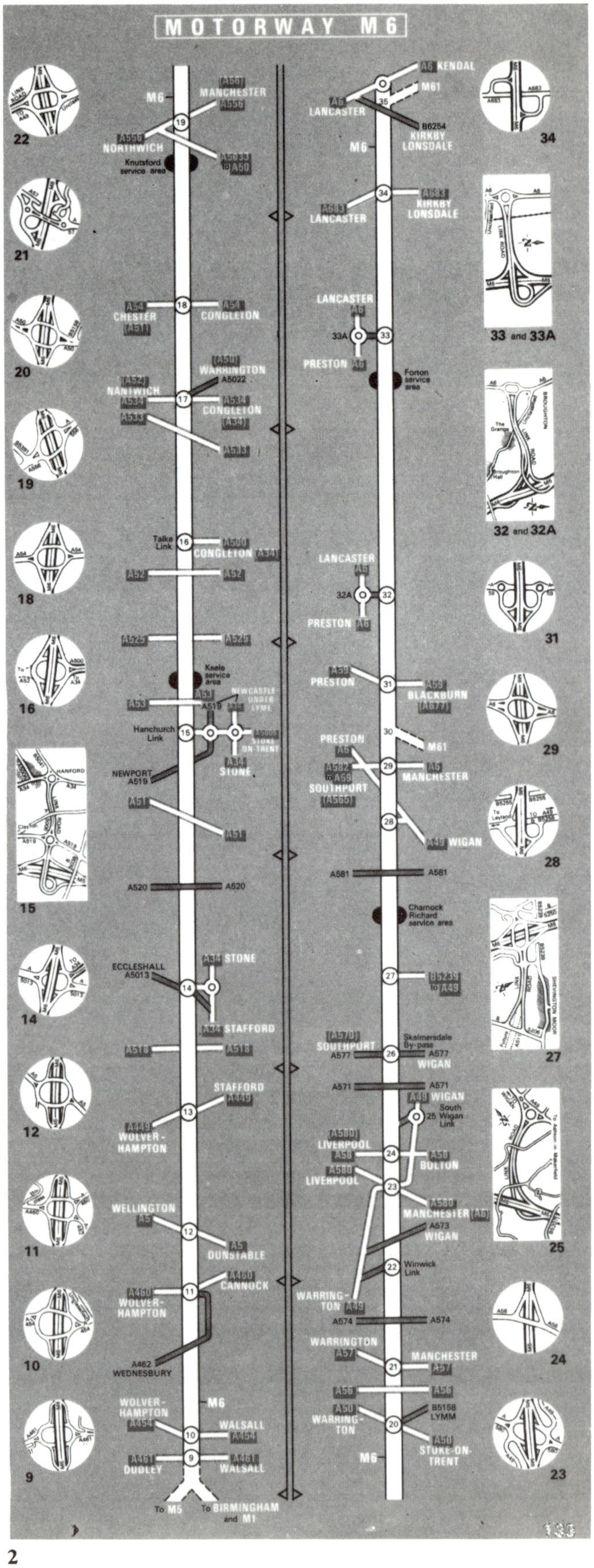

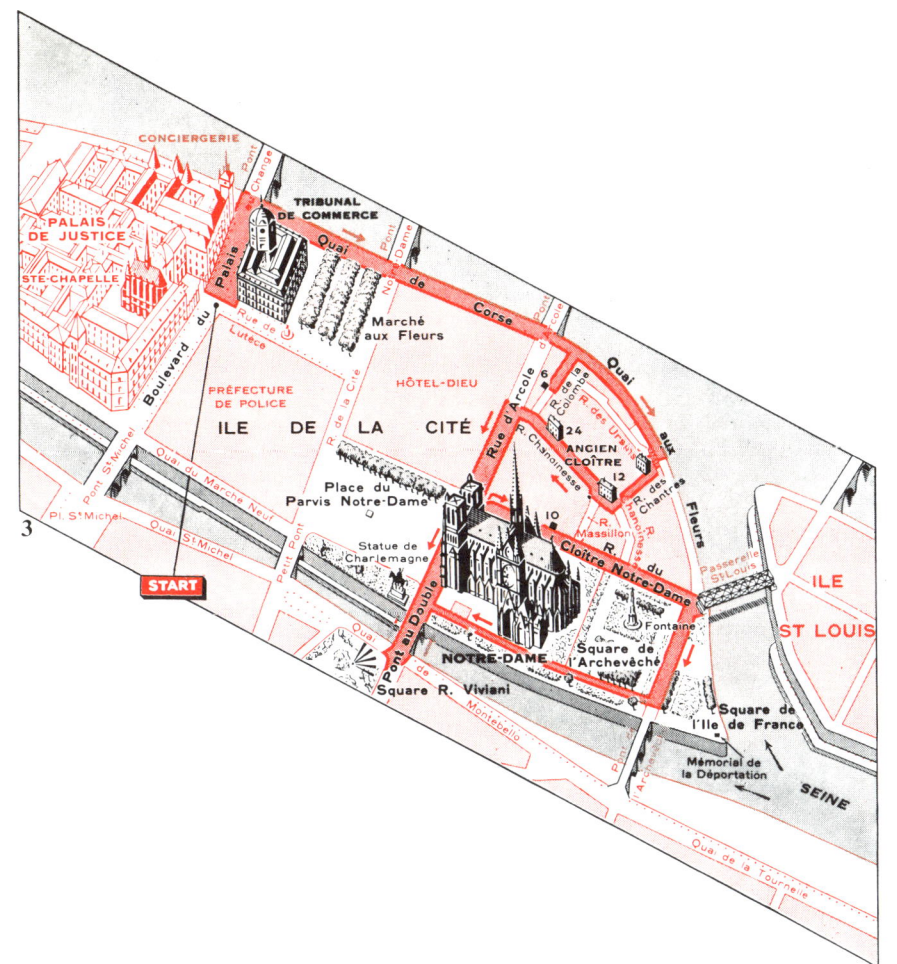

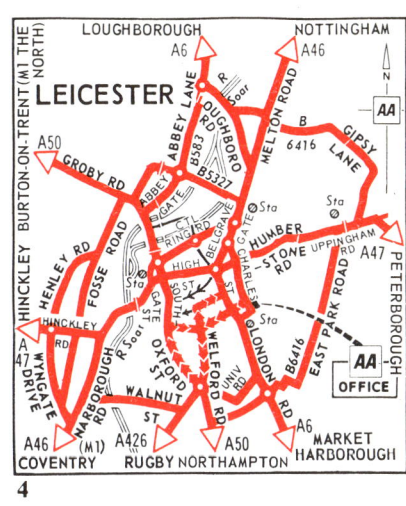

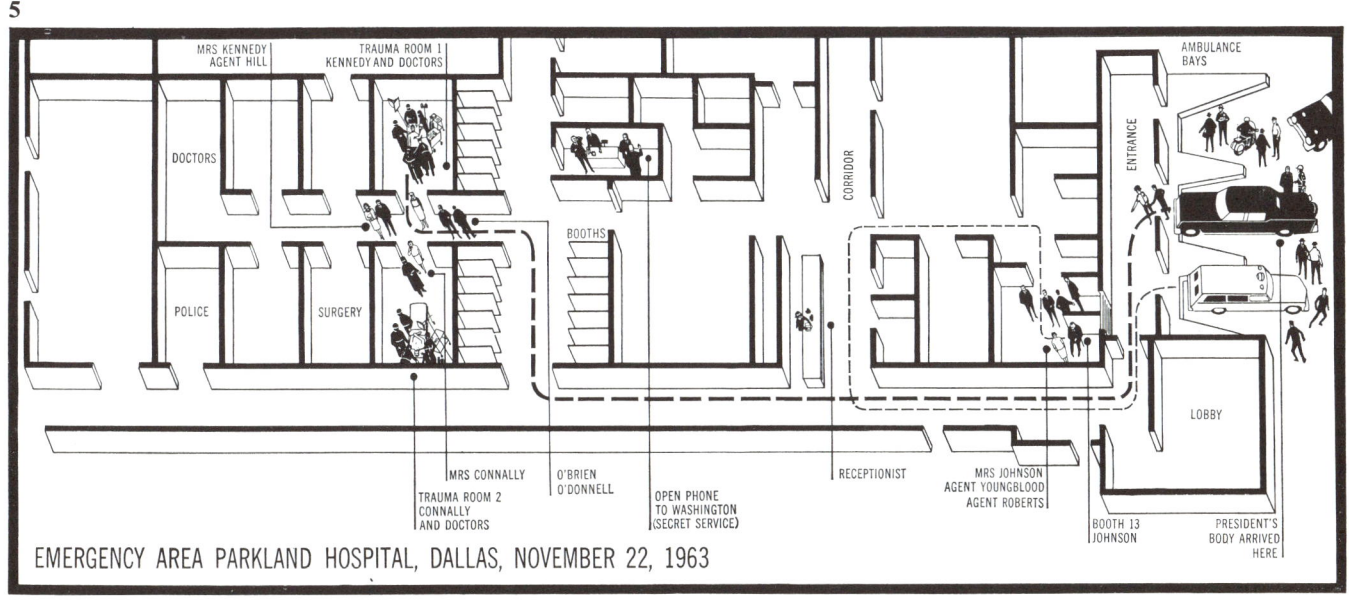

EMERGENCY AREA PARKLAND HOSPITAL, DALLAS, NOVEMBER 22, 1963

Route map

1 Map from an atlas, showing the movements of 'Beaker and Battle-axe people'. Where routes are imprecise, broad sweeping arrows or, as in this map, many small arrows, solve the lack of definite information. *Complete Atlas of the British Isles,* Reader's Digest

2 Map from a book, showing the three voyages of Captain Cook. Route on white ground; all labelling omitted; result, a clear, efficient comparison of routes. Geographical Projects map. Frank Debenham, *Discovery and Exploration,* Paul Hamlyn

3 Part of a map from a book, showing the exploration of the Middle East by Alexander the Great. The relief base shows clearly why certain routes were taken. One problem in using relief model maps, that of labelling in dark shadow areas, has been avoided. The routes are well coded (the longest and most complex has the simple line, shorter and straighter routes have a complex line), arrows give direction, numbers give the date of the particular journey and avoid a more elaborate system of lines, latitude and longitude are included and the scale is clear. Original in two colours. Geographical Projects map. Frank Debenham, *Discovery and Exploration,* Paul Hamlyn

4 Map from a book, showing the voyages of Columbus. Using the globe has impact, and the routes are clearly coded. But the important information is not clear in the West Indies, where the perspective distorts the map. Geographical Projects map. Frank Debenham, *Discovery and Exploration,* Paul Hamlyn

1

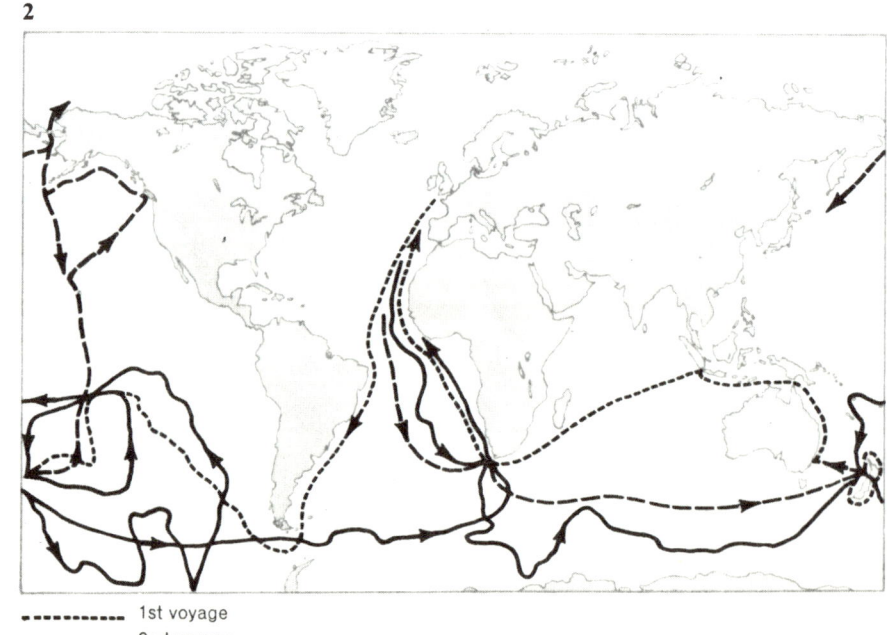

2

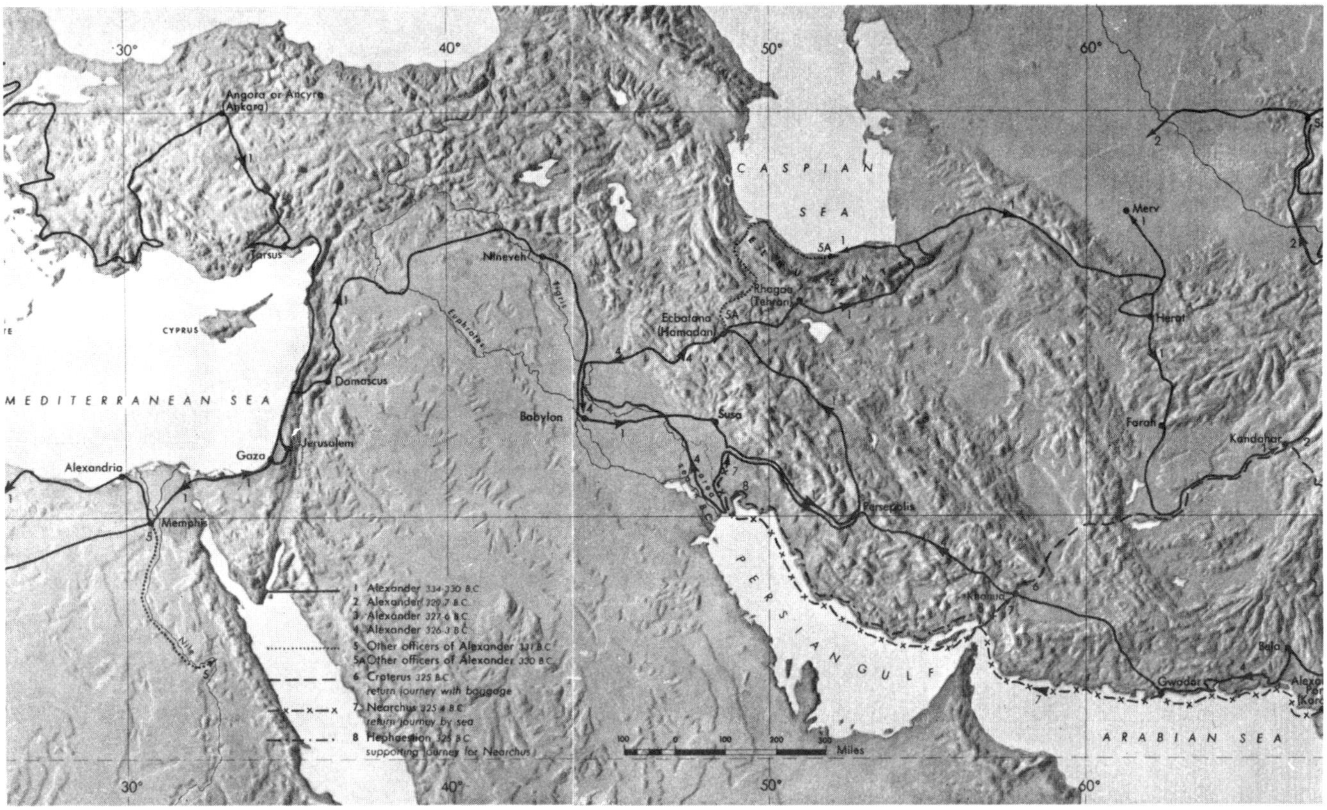

3

4

The voyages of Columbus.

— · — · — Columbus' 1st voyage 1492-3
— — — — Columbus' 2nd voyage 1493-6
· · · · · · · · · Columbus' 3rd voyage 1498
——————— Columbus' 4th voyage 1502-4

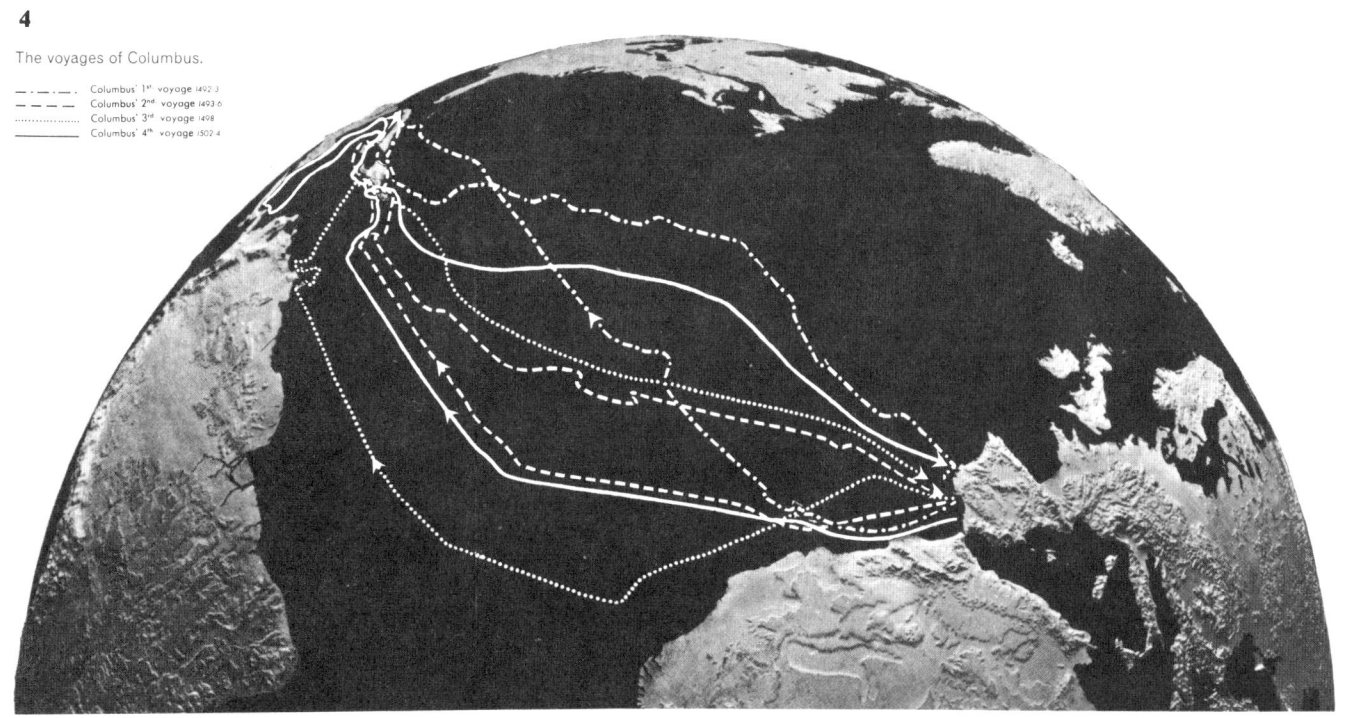

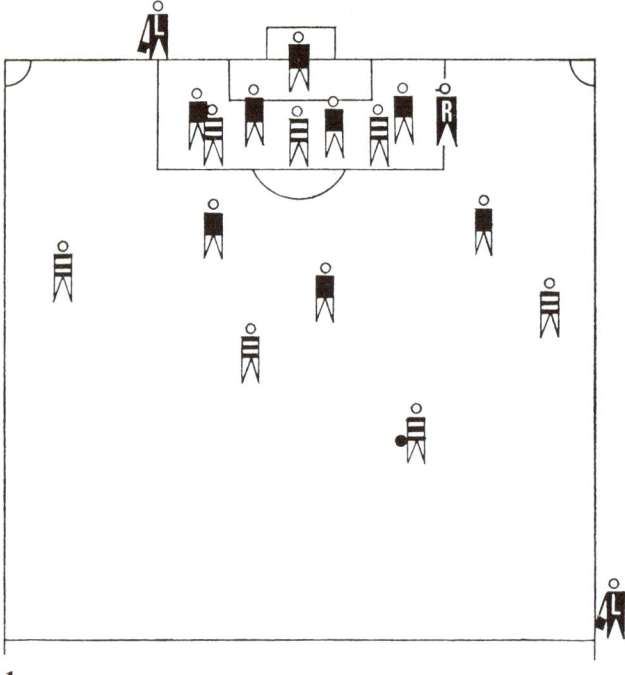

1

2

3

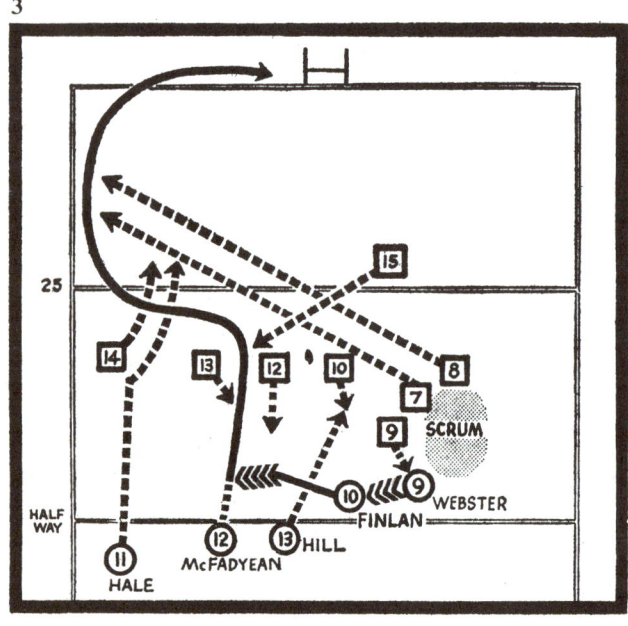

Games and battles

1 Diagram from a book, showing free-kick close to goal in a football match. The essential thing is to recognize the two sides, referee and linesman. Dennis Howell, *Soccer Refereeing,* Pelham Books

2 Diagram from a book, showing the situation of the ball hitting the referee. Pin men have their uses: they can show 'movement' and are more human than Isotype men. *Know the Game, Association Football,* Educational Productions

3 Diagram from a newspaper, showing how a try was scored in a rugby match. A typical diagram showing clearly the movement of players. *Sports Argus*

4 Diagram from a magazine, showing the battle of Coronel. Models are used to show the battle. The situation is that of British ships coming out of the sunset afterglow to encounter a German squadron, hidden by the failing light. Darkness sets in and two British ships are sunk while the others escape. *History of the 20th Century,* Purnell

5 Map from a paperback history atlas, showing the German attack on France 1914–18. Such maps are used to show the movement of armies and the changing front line over a period of time – in this case four years. *dtv-Atlas zur Weltgeschichte,* Deutscher Taschenbuch Verlag

6 Map from a school textbook, showing the same situation simplified for children. A. M. Newth, *Britain in the Modern World,* Penguin Books

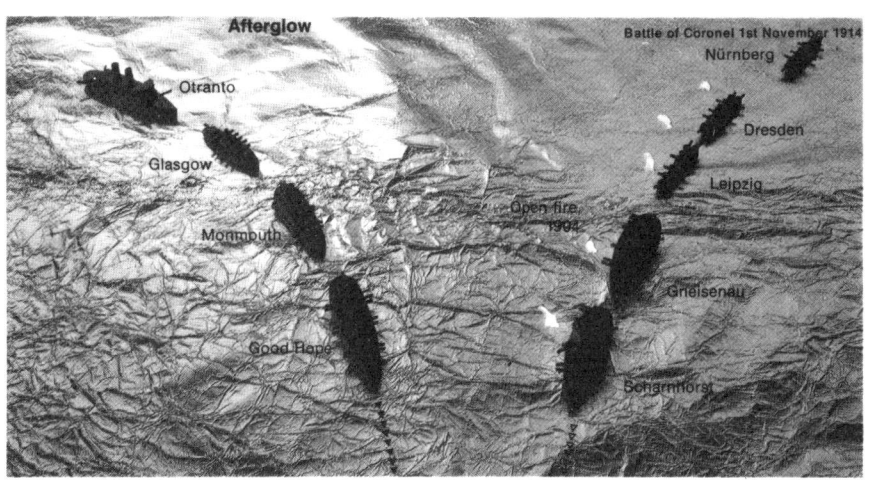
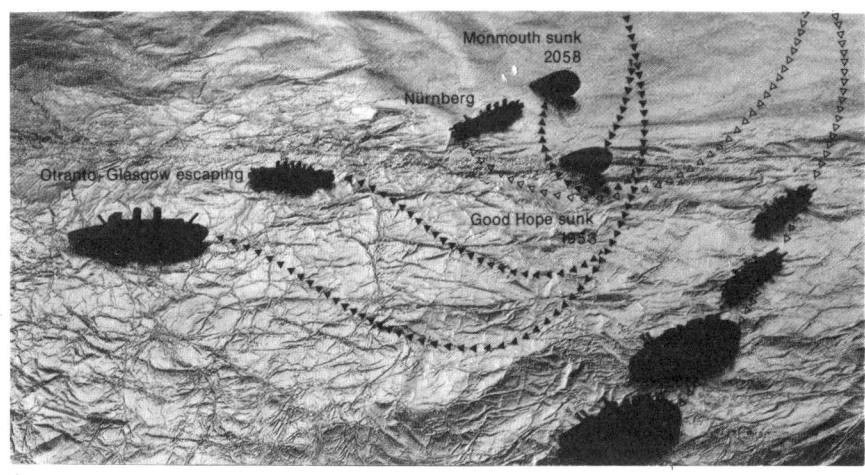

4

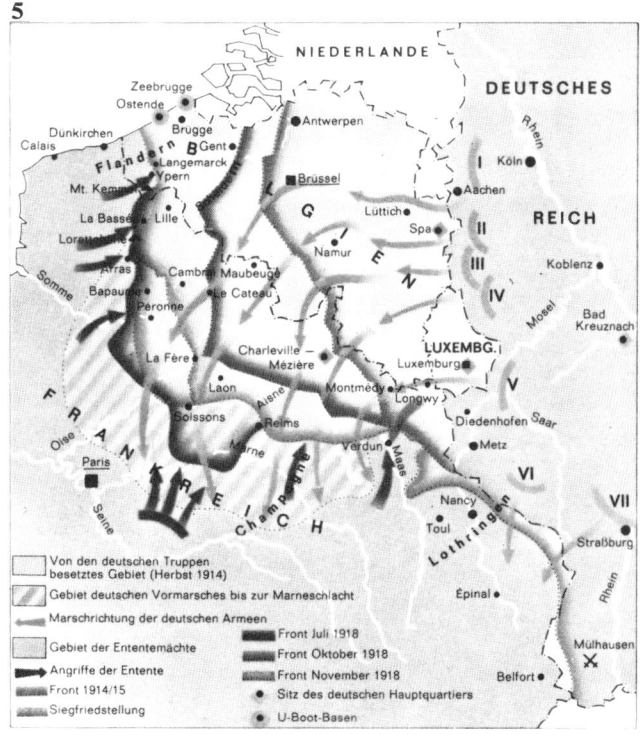

5

Von den deutschen Truppen besetztes Gebiet (Herbst 1914)
Gebiet deutschen Vormarsches bis zur Marneschlacht
Marschrichtung der deutschen Armeen
Gebiet der Ententemächte
Angriffe der Entente
Front 1914/15
Siegfriedstellung
Front Juli 1918
Front Oktober 1918
Front November 1918
Sitz des deutschen Hauptquartiers
U-Boot-Basen

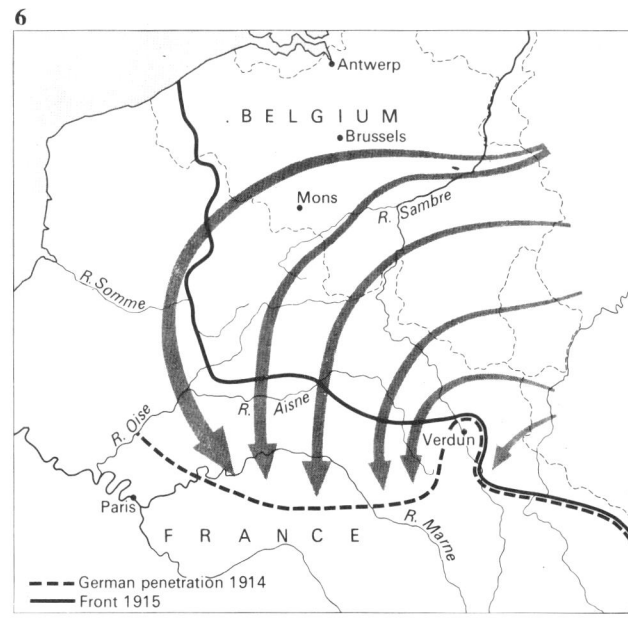

6

German penetration 1914
Front 1915

Colour

1 Map showing United States. Political, non-quantitative. In colouring any such map four colours are sufficient to differentiate each county or country from its neighbours.

2 Part of a group of maps from an atlas showing distribution of religious groups in the British Isles. In making a quantitative statement it is important that each group is clearly shown and an understandable pattern of colouring is created. Dark for large percentages, bright for middle percentages, light or grey for low percentages. *Complete Atlas of the British Isles,* Reader's Digest

3 Map from a magazine, showing linguistic division in Austria-Hungary and the Balkans. The key lists nineteen groups. The complex mix-up of groups is shown on the map. *History of the 20th Century,* Purnell

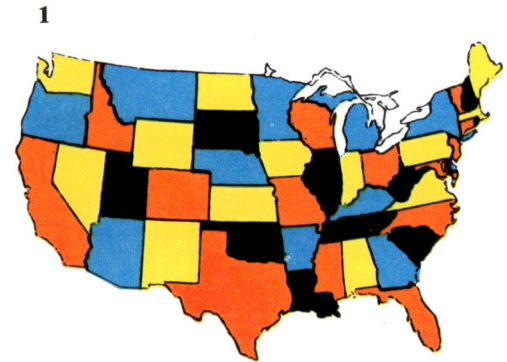

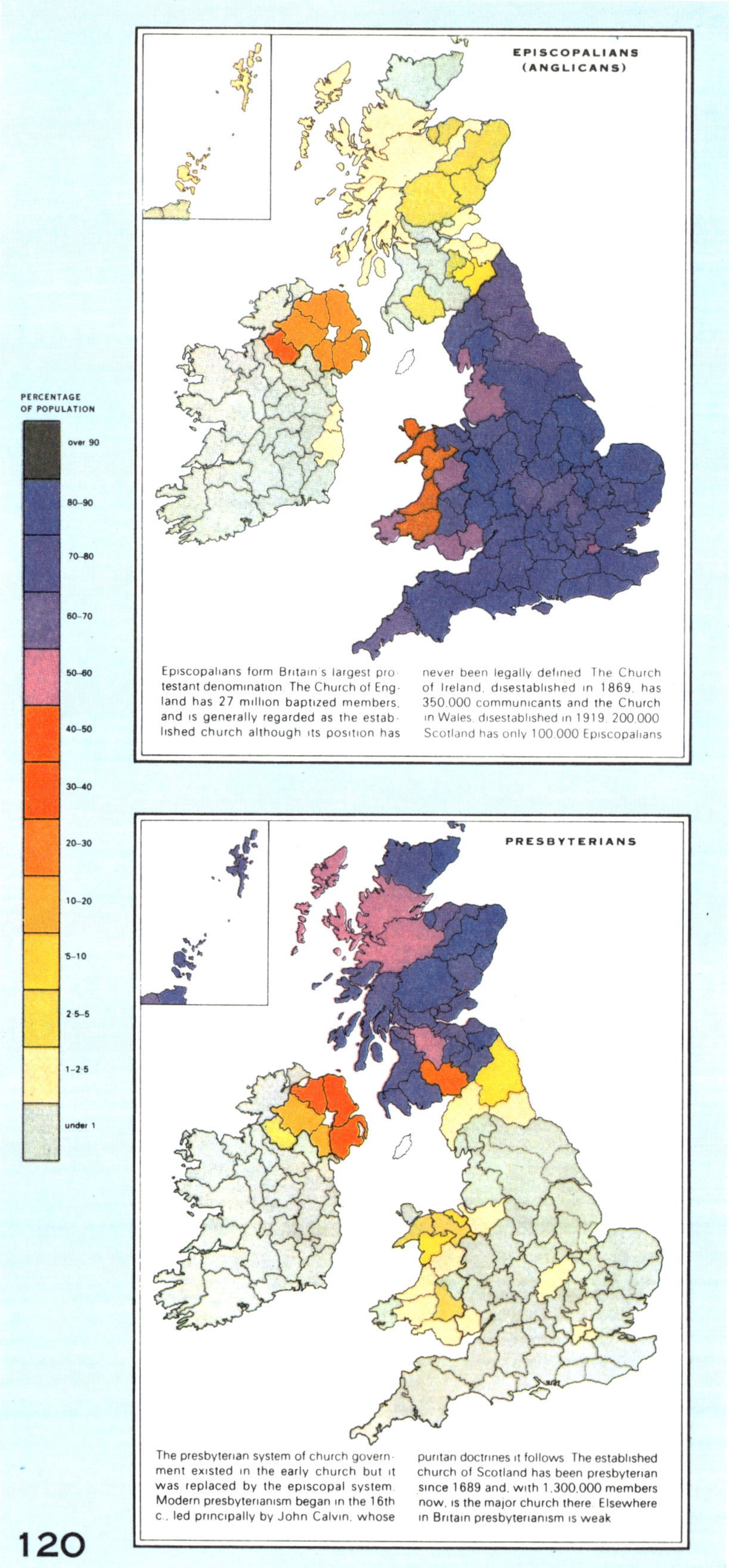

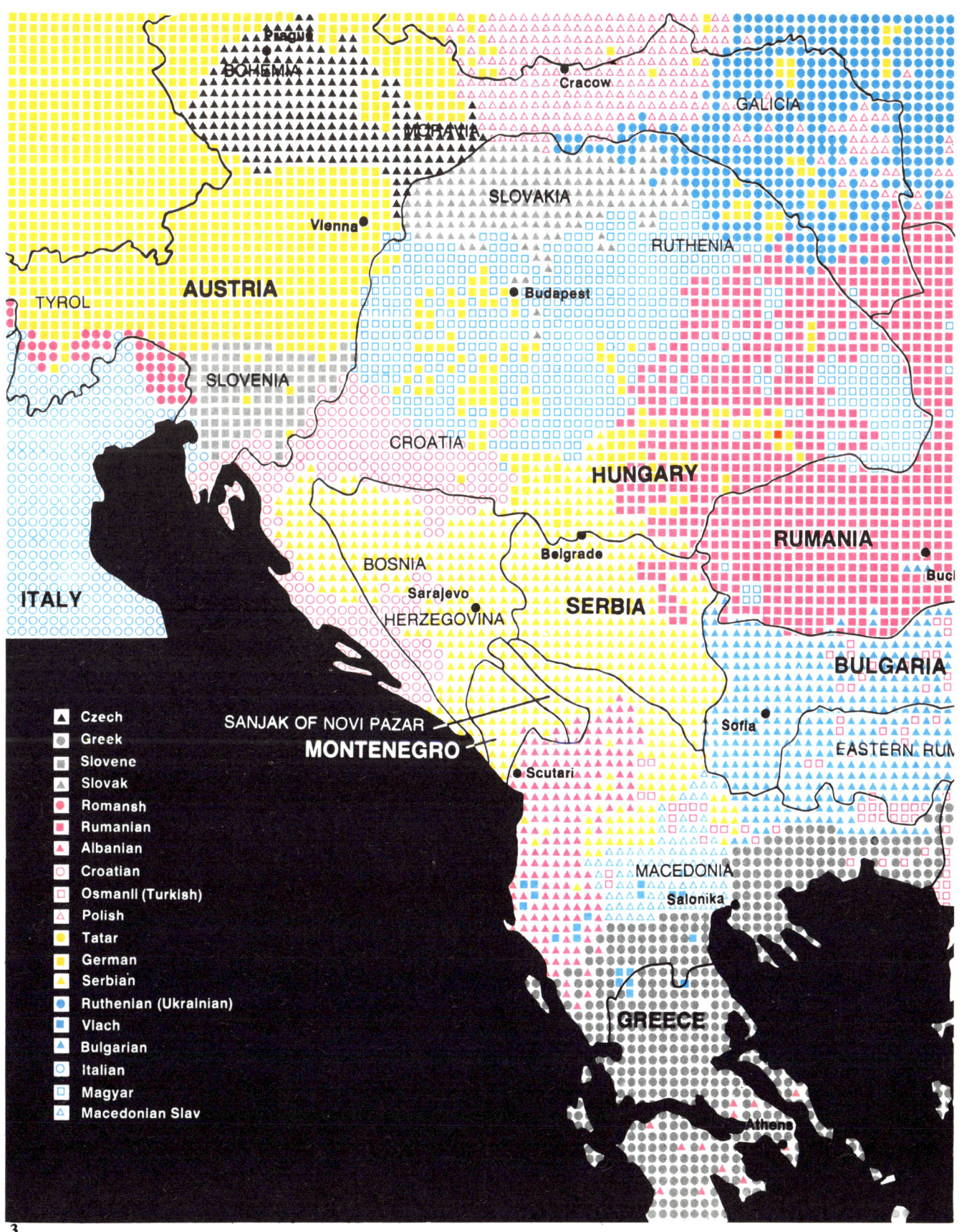

Texture and shading

1 Map from an atlas, showing rainfall for Africa. A quantitative statement.

2 Map from an atlas, showing natural vegetation for Africa. A non-quantitative statement. These two maps were adjacent in the atlas. There were other rainfall and vegetation maps for the main land masses. From the carefully chosen textures, the subject of each map is quite clear. Texture attempts to show the nature of the vegetation. On these two maps twelve completely different textures are used. *Knaurs Welt-Atlas*

3 Map from a paperback history atlas, showing spread of Islam. Texture can limit the amount of labelling that can be included. If texture is applied in broad bands of light tone, all the advantages of a variety of textures can be obtained and complex labelling included. Edges of the bands can fade to indicate indefinite edge– something that very few line textures can show satisfactorily. *dtv-Atlas zur Weltgeschichte,* Deutscher Taschenbuch Verlag

4 Map from a text book, showing population density in urban local government areas. Proportional line shading to show statistics. G. C. Dickinson, *Statistical Maping and the Presentation of Statistics*, Edward Arnold

5 Some of the range of textures in Zip-a-tone, one of the methods of applying mechanical textures to maps. Contrast with the range possible when only tone is used.

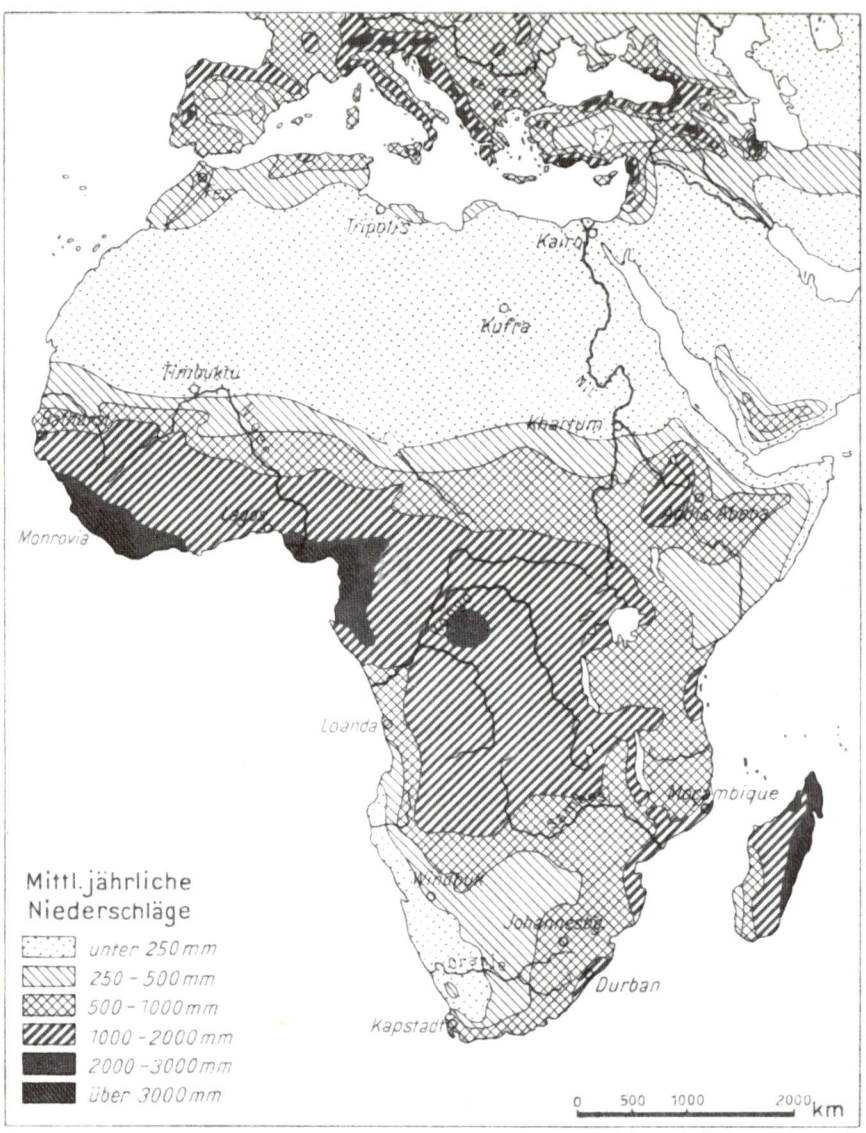

1

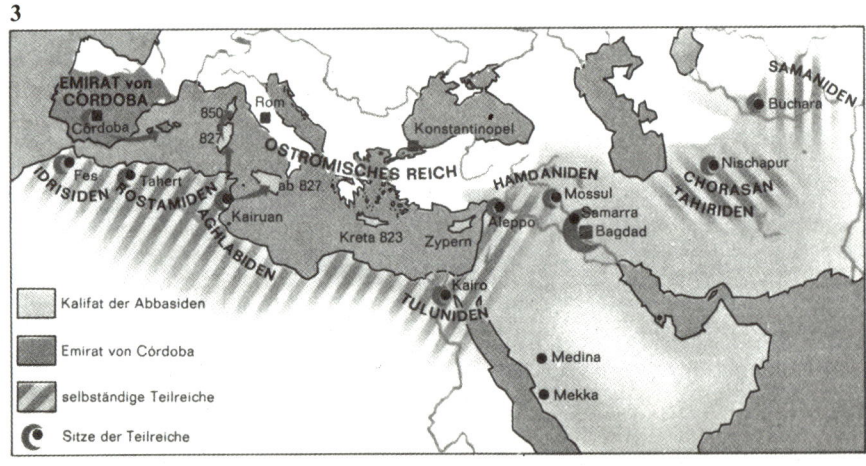

3

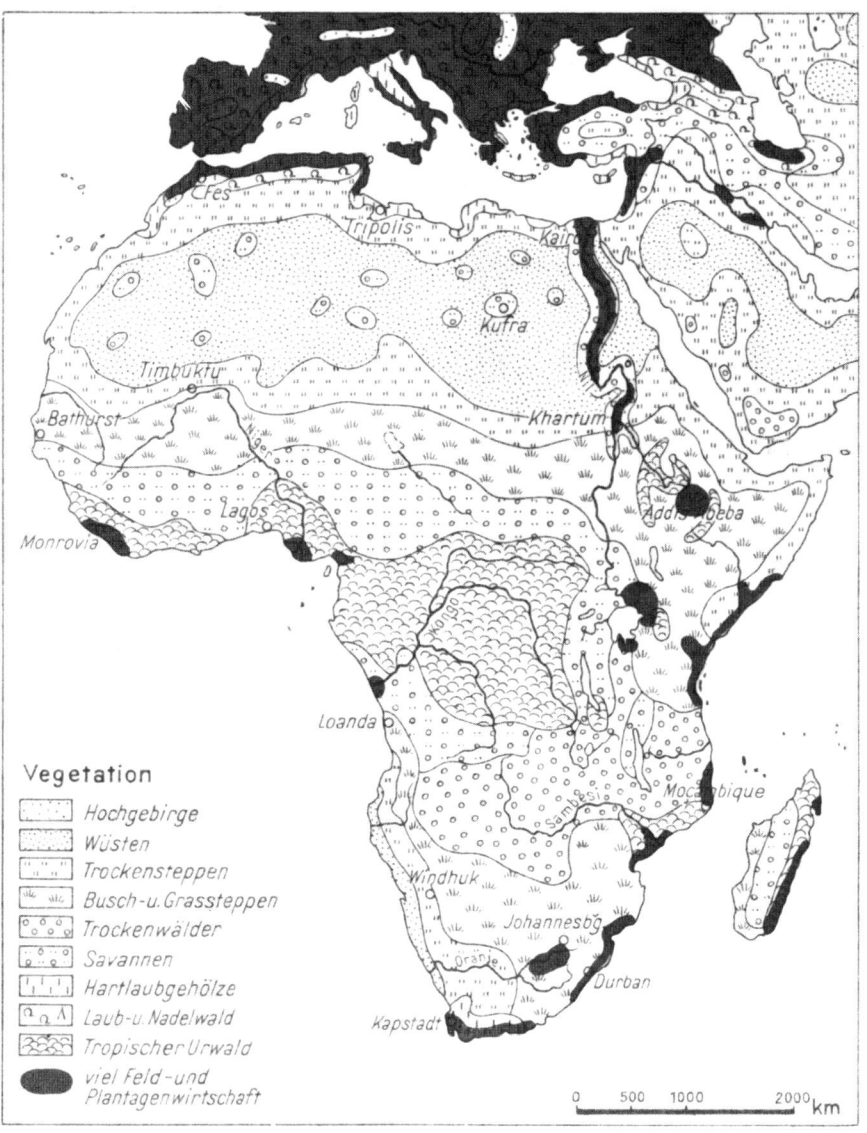

2

Vegetation
- Hochgebirge
- Wüsten
- Trockensteppen
- Busch- u. Grassteppen
- Trockenwälder
- Savannen
- Hartlaubgehölze
- Laub- u. Nadelwald
- Tropischer Urwald
- viel Feld- und Plantagenwirtschaft

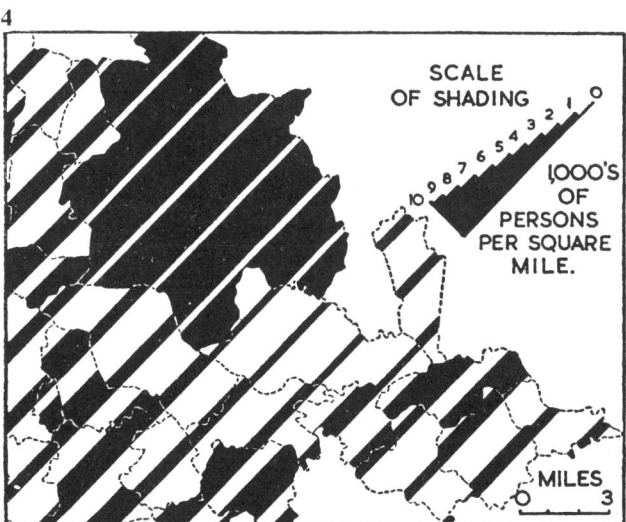

4

SCALE OF SHADING
10 9 8 7 6 5 4 3 2 1 0
1,000'S OF PERSONS PER SQUARE MILE.

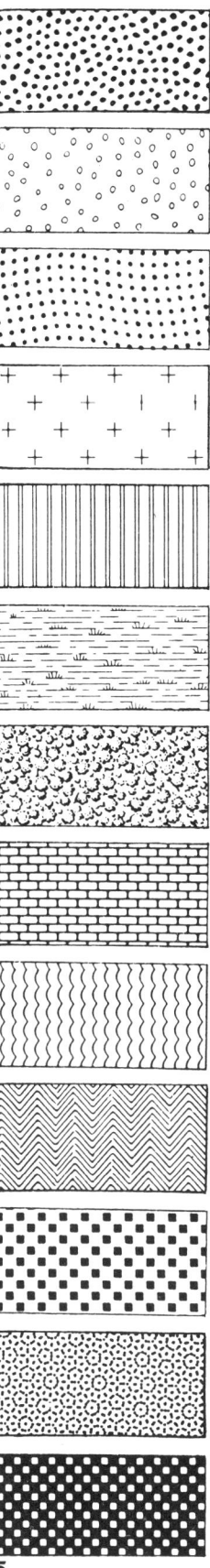

5

- ■ WHEEL
- ■ MODEL WHEEL
- ▲ BURIAL OF PAIRED OXEN
- ▲ MODEL OF PAIRED OXEN
- ◉ MODEL WAGON
- ● CART

1

Line and tone

1 Map from a scientific magazine, showing presence of wagons in Europe before 2000BC. Hachures are lines drawn down the slope in the direction of the steepest gradient; they are closer together where the slope is steeper. Elegantly done, they show the physical features and in no way interfere with reading the symbols. Original in two colours. *Scientific American*

2 Map, showing contour lines, with grey shadow added to emphasise the relief features. This produces a step-like appearance, but is more suggestive of height than contour lines by themselves.

3 Map from a textbook, showing two flat tones representing higher ground. A simple solution if the map is to be reproduced small.

4 Map showing tone being used to model physical features.

5 Map from an atlas, showing the number of days when fog occurred in the British Isles. A quantitative statement. There is a definite limit to the number of tones which can be read clearly. Lines round the edges help to differentiate the areas. *Complete Atlas of the British Isles*, Reader's Digest

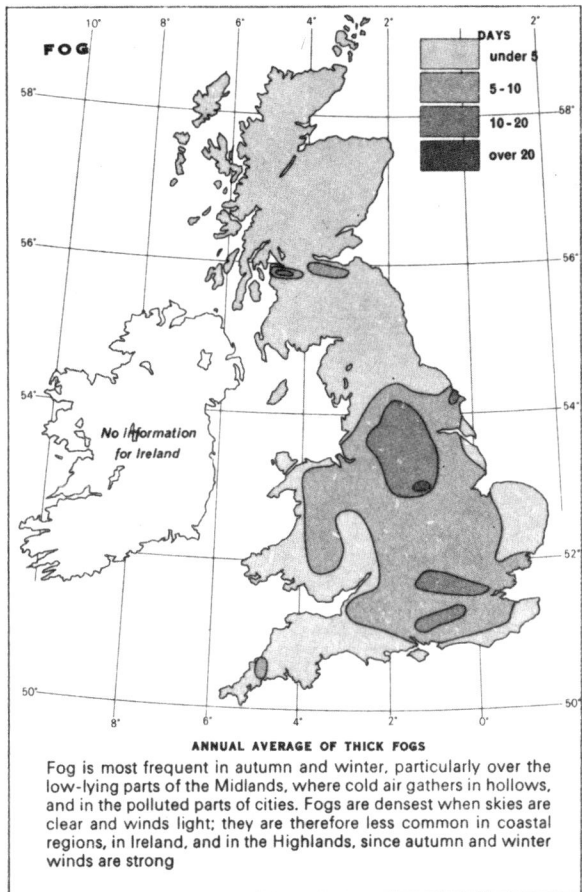

5

ANNUAL AVERAGE OF THICK FOGS
Fog is most frequent in autumn and winter, particularly over the low-lying parts of the Midlands, where cold air gathers in hollows, and in the polluted parts of cities. Fogs are densest when skies are clear and winds light; they are therefore less common in coastal regions, in Ireland, and in the Highlands, since autumn and winter winds are strong

2

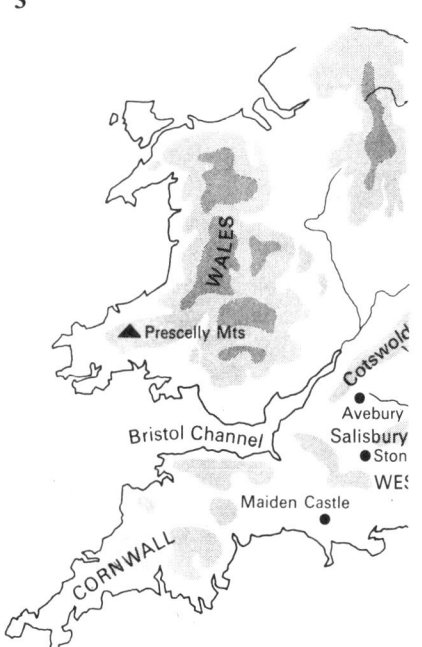

3 4

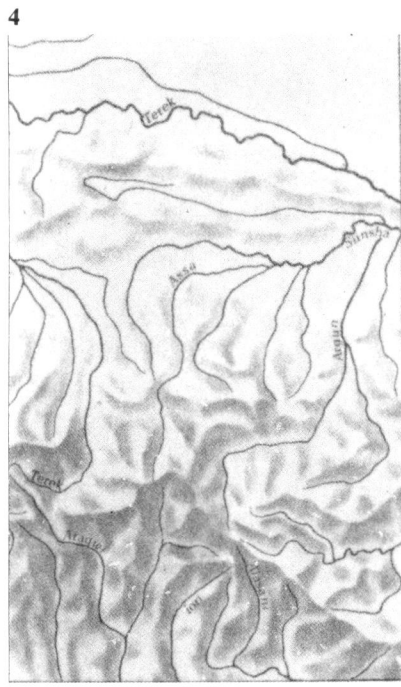

83

Bars

1 Map from a school textbook, showing the most important harbours and their annual turnover in goods (in million tons). Bars located at points (see page 86). Unfortunately the New York bar was too long to fit. A. Jentzsch and J. Winkler, *Länderkunde,* Westermann

2 Map from a history atlas, showing immigration to the United States. Lining bars along the same base enables figures to be read easily. *Atlas zur Weltgeschichte,* Westermann

3 Map from a government publication, showing world trade in wheat. Base lines, with no quantities shown (Canada, New Zealand), dispel any doubt in the reader's mind that something has been left off. *A Graphic Summary of World Agriculture,* US Department of Agriculture

4 Map from a government publication, showing the changing pattern of population distribution 1901–61 in London. It is only possible really to compare complicated groups of bars when they have a common base line, and when the map element is not competitive. Ministry of Housing and Local Government

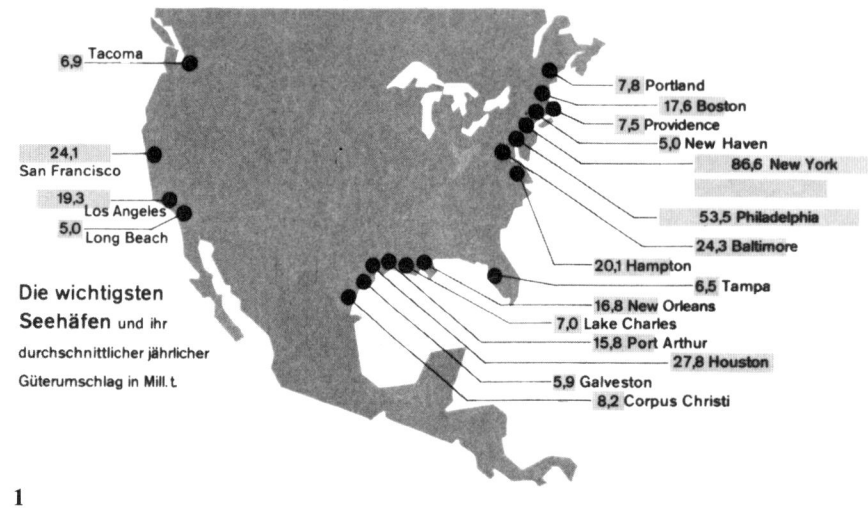

1

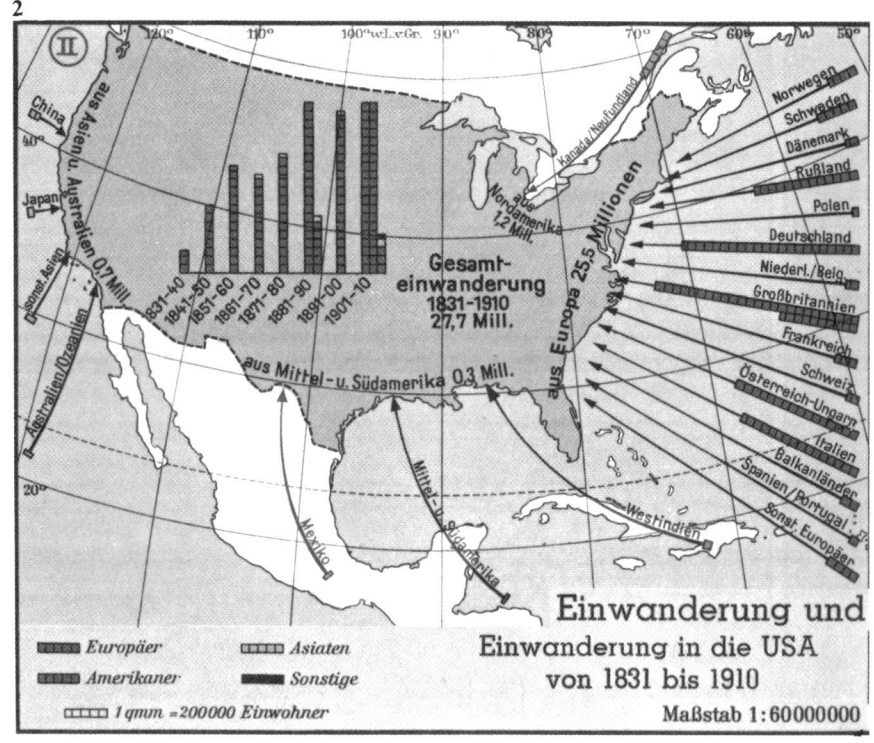

3

World Trade in Wheat, Average 1957-61*

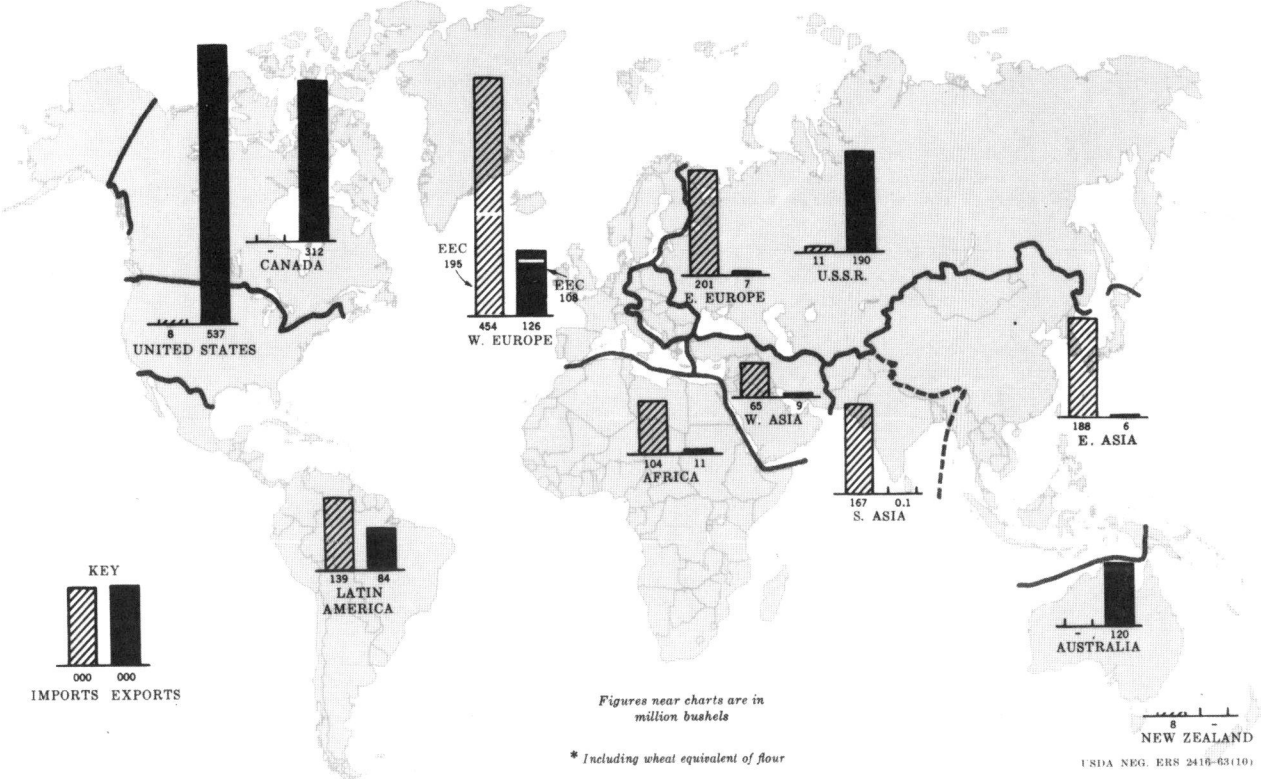

Figures near charts are in million bushels

* Including wheat equivalent of flour

USDA NEG. ERS 2416-63(10)

4

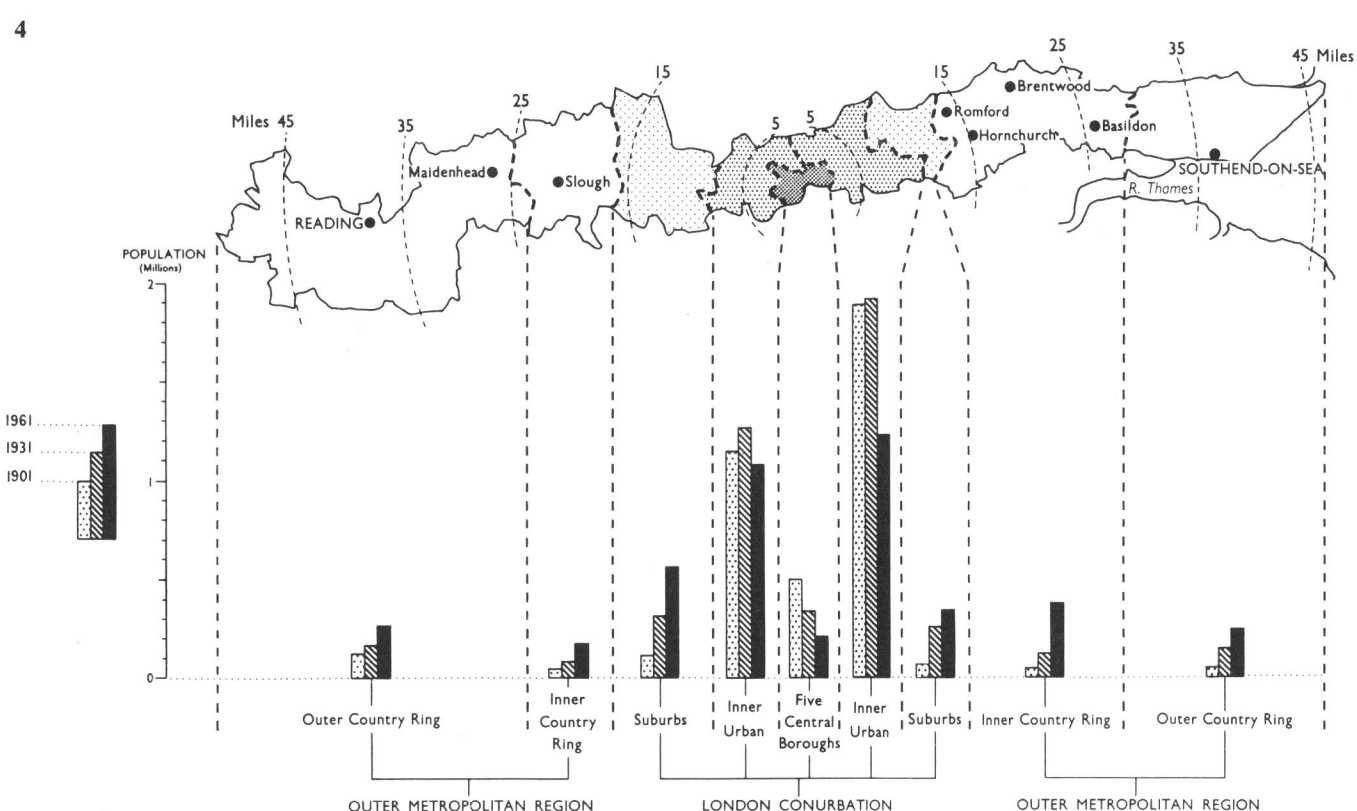

Blocks

1 Map from textbook, showing located block-piles, representing freight handled by ports in Western Europe for four selected years. F. J. Monkhouse and H. R. Wilkinson, *Maps and Diagrams,* Methuen

2 Map from a book, showing United States university libraries and the number of volumes they contain. The problem of severe crowding in the foreground has been neatly overcome by using exaggerated perspective. K. D. Metcalf, *A Graphic Summary,* Harvard University Library

3 Map from a magazine, showing the balance of power in 1914. Problems of reading the information are compensated by symbols and complete statistics tabulated in the chart below the map. *History of the 20th Century,* Purnell

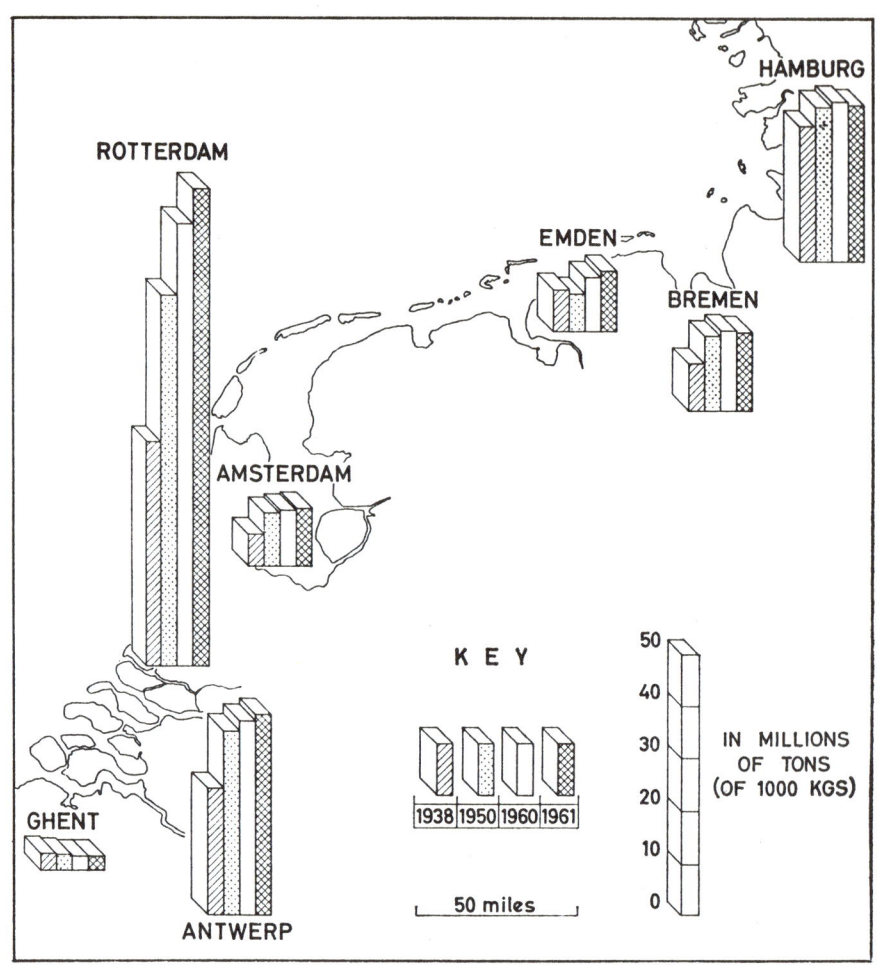

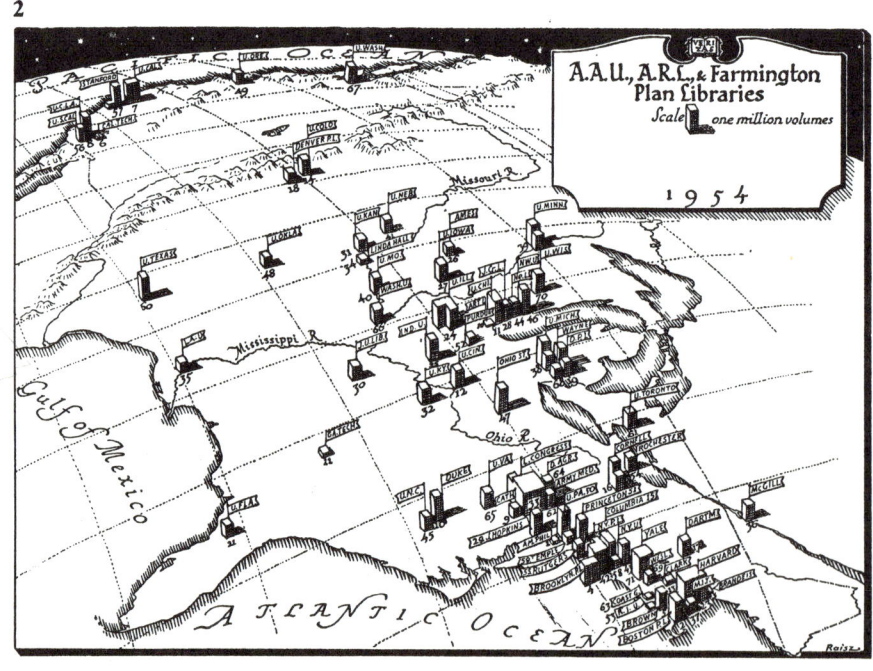

The Balance of Power, 1914

	Great Britain	France	Russia	Germany	Austria-Hungary	Turkey
Population	46,407,037	39,601,509	167,000,000	65,000,000	49,882,231	21,373,900
Soldiers available on mobilization	711,000	3,500,000	4,423,000	8,500,000	3,000,000	360,000
Merchant fleet (net steam tonnage)	11,538,000	1,098,000	(1913) 486,914	3,096,000	(1912) 559,784	(1911) 66,878
Battleships (built and being built)	64	28	16	40	16	
Cruisers	121	34	14	57	12	
Submarines	64	73	29	23	6	
Annual value of foreign trade (£)	1,223,152,000	424,000,000	190,247,000	1,030,380,000	198,712,000	67,472,000
Annual steel production (tons)	6,903,000	4,333,000	4,416,000	17,024,000	2,642,000	
Railway mileage	23,441	25,471	46,573	39,439	27,545	3,882

¹ Including empire ² Immediate mobilization ³ Emergency maximum

1
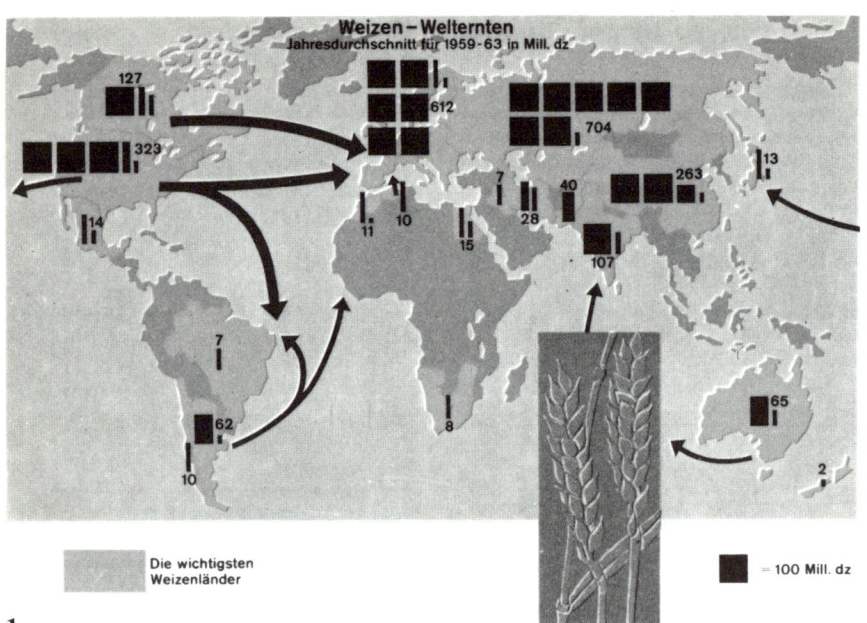

2
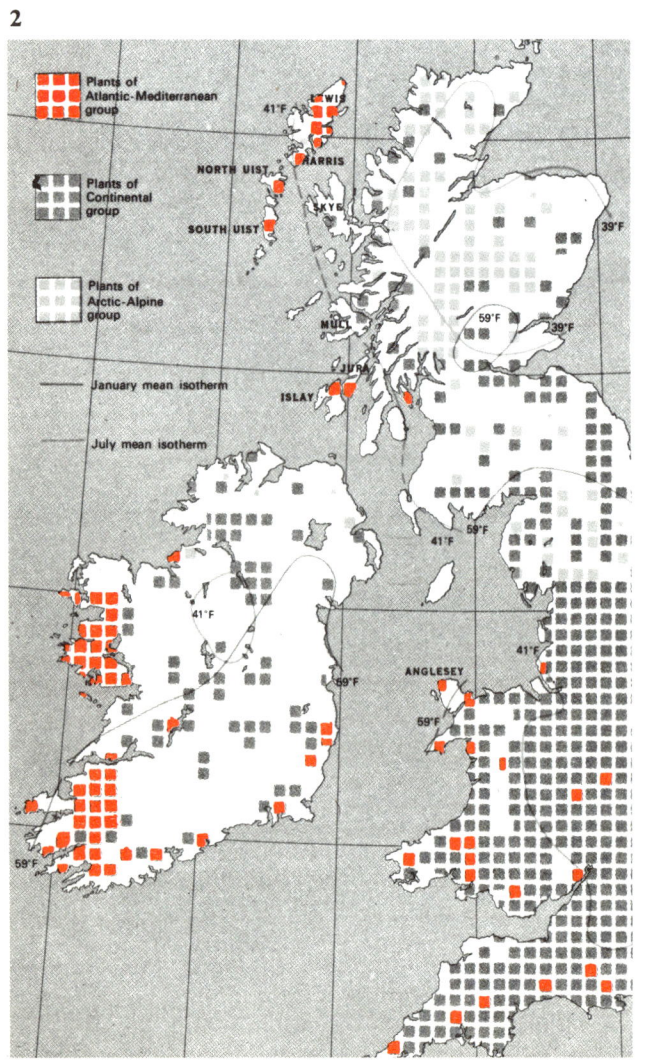

3
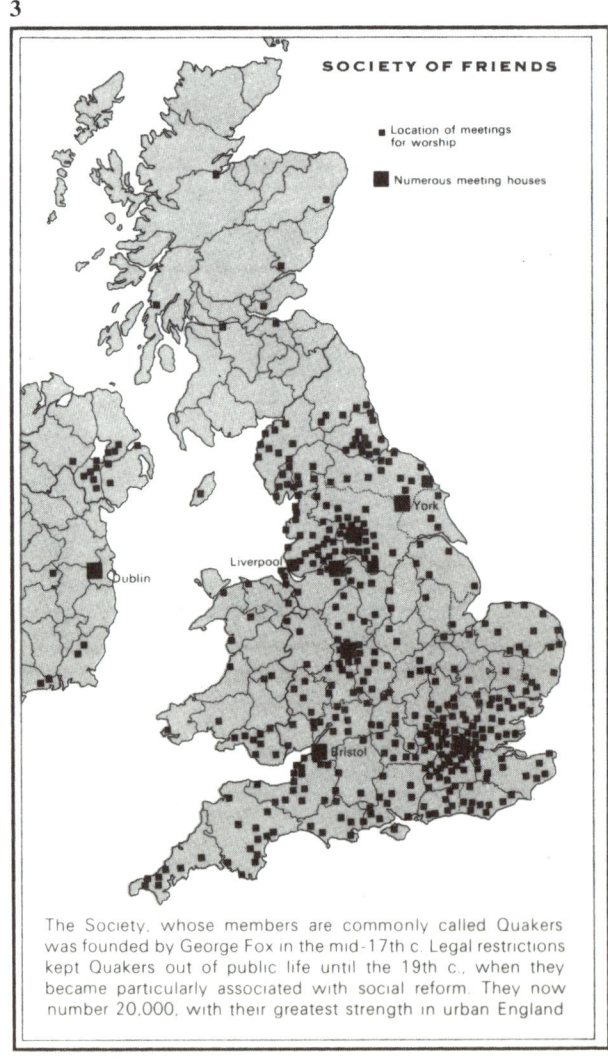

The Society, whose members are commonly called Quakers was founded by George Fox in the mid-17th c. Legal restrictions kept Quakers out of public life until the 19th c., when they became particularly associated with social reform. They now number 20,000, with their greatest strength in urban England

Squares

1 Map from a school textbook, showing world production of wheat (see page 104). Squares are used as units of bars superimposed on the map. This technique overcomes problems of comparing different lengths of bars. A. Jentzsch and J. Winkler, *Güterkunde*, Westermann

2 Map from an atlas, showing distribution of three plant groups. Original in colour. A non-quantitative statement, where squares are used so that an intermingled distribution can be seen. *Complete Atlas of the British Isles*, Reader's Digest

3 Map from an atlas, showing location of meetings and meeting houses of the Society of Friends. *Complete Atlas of the British Isles*, Reader's Digest

4 Map from a history atlas, showing population displacement as a result of World War II. Squares proportional to quantities. Original in full colour. *Atlas zur Weltgeschichte*, Westermann

5 Map from a paperback history atlas, showing Germany after World War I. Proportional squares superimposed show the decrease in Germany's size and population as a result of boundary changes. *dtv-Atlas zur Weltgeschichte*, Deutscher Taschenbuch Verlag

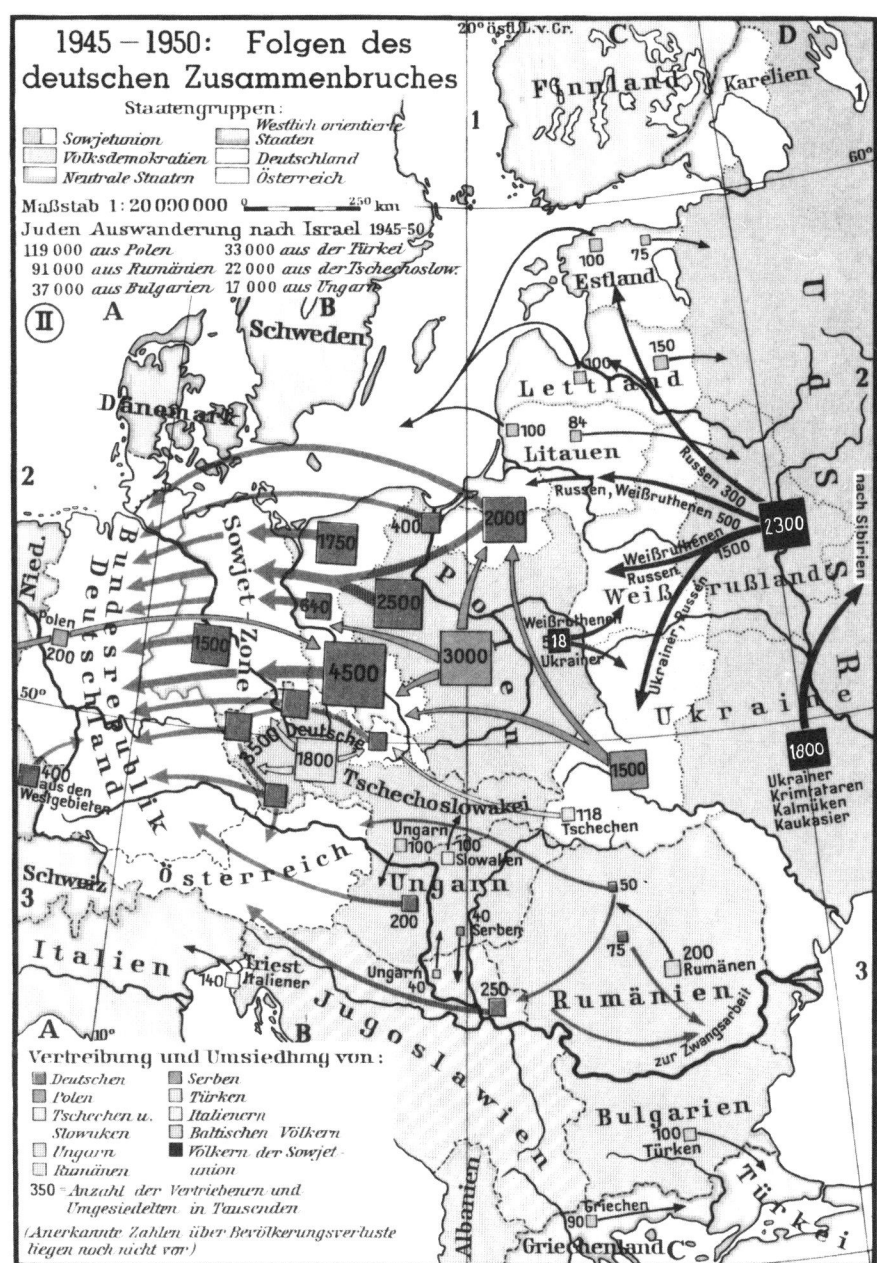

5

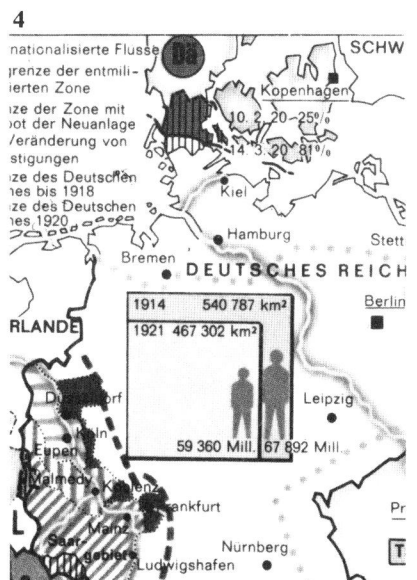

4

Some European Countries, 1960
Population   Densities

Each dot represents 1 million people

1

Each dot represents
10 people per square kilometer

Dots and circles

1   Map from a school textbook, showing population and densities of some European countries. A simple diagram which shows the use of dots as overall texture to fill an area. There is no grouping of dots to show density of large cities. Because some areas are small, Belgium's density appears greater. Diagram on left shows that it is the same. Marie Neurath, *Living with One Another,* Parrish

2   Map from an atlas, showing population density. Compare this treatment of Europe with that of previous diagram. *March of Civilisation,* Hammond

3   One of a series of maps from a magazine, showing the probable representation of 5th Republic deputies in each department of France. Circles are in proportion to the estimated numerical representation, size of the circle indicates the importance of the gain. This map was produced electronically. (A whole series of maps was used inside.) Proportional circles texturing electoral areas of France. *Paris Match*

2

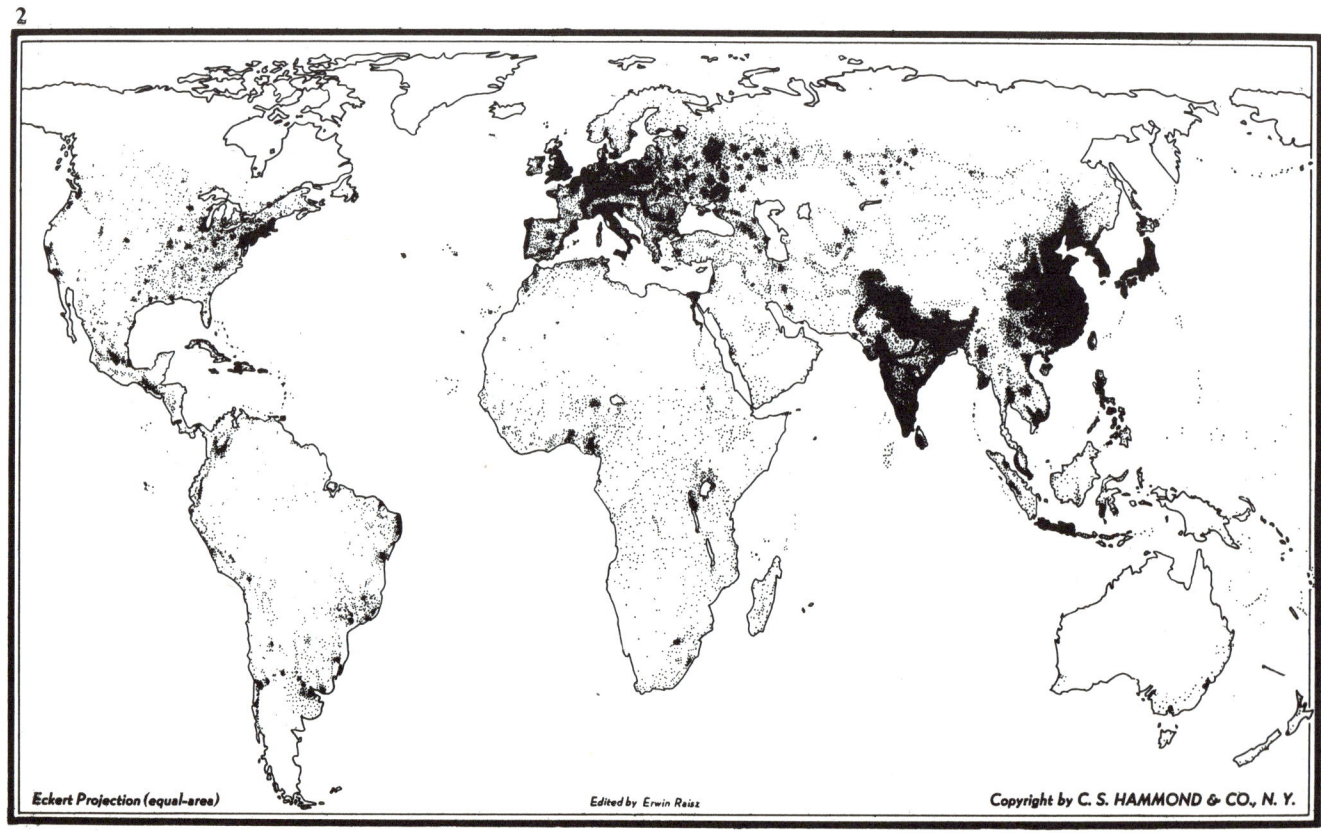

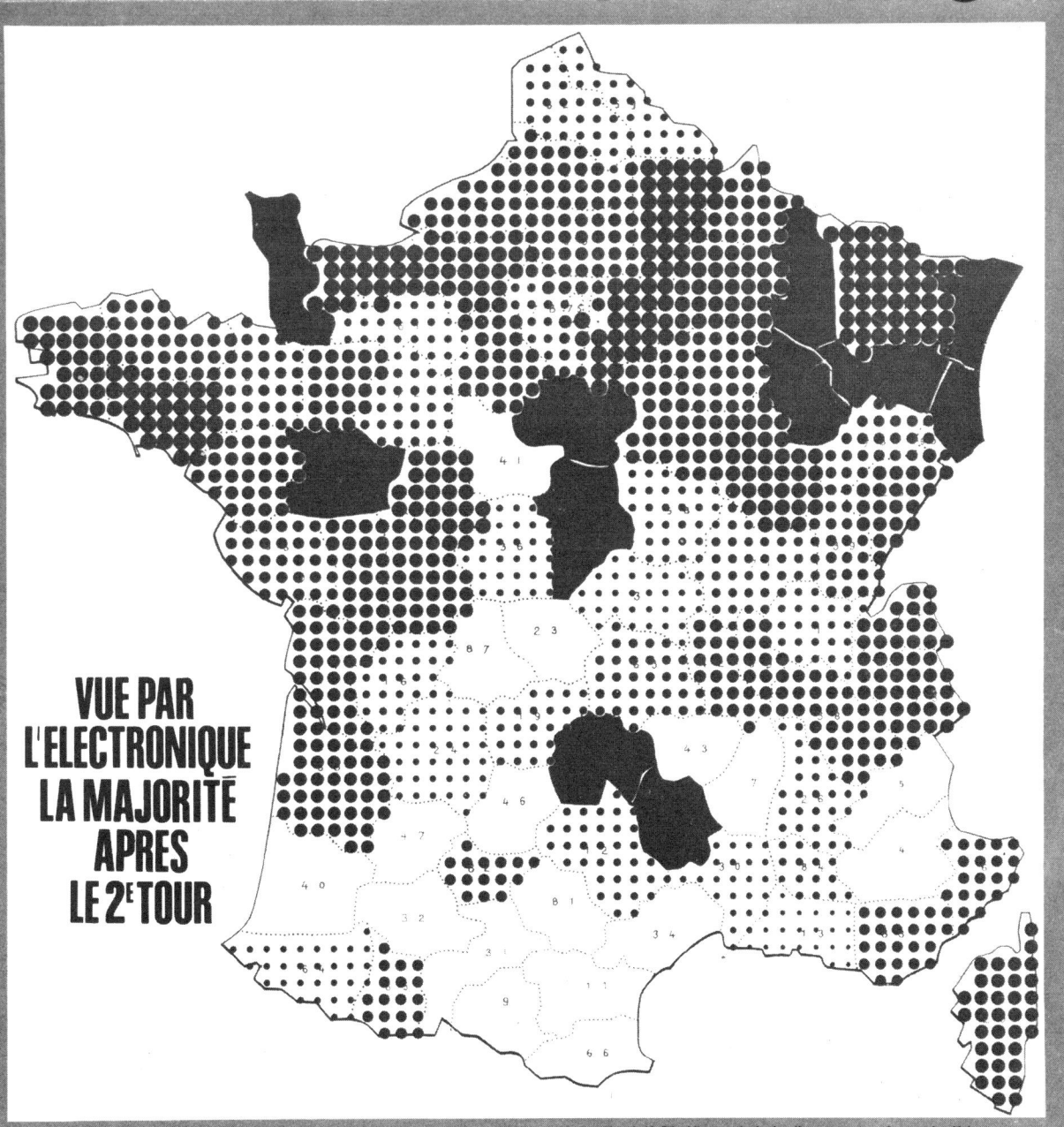

Circles and divided circles

1 Map from a paperback history atlas, showing religious situation in the late 17th century. Circles show Protestant, Roman Catholic and Islam (sphere) countries and influence. Original colour coded. An explanatory map. *dtv-Atlas zur Weltgeschichte,* Deutscher Taschenbuch Verlag

2 Map from an atlas, showing catches of fish. Open circle overcomes confusion caused by close overlapping of circles. But the problem of sorting out seven sizes of circle remains. *Shorter Oxford Economic Atlas,* Oxford University Press

3 Map from an atlas, using divided circle. The smallest size at which a divided circle will be clear has dictated the scale of circles and caused the congestion on Europe.

4 Map from a text book, showing import of wheat as a percentage of total need. Overlapping problem solved. A. Jentzsch and J. Winkler, *Länderkunde,* Westermann

5 Part of a wall map of Africa, showing population, white and non-white. Getting the largest circle to fit into Nigeria dictates the proportions of the other circles. Good labelling overcomes comparing sizes of circle and sectors. *Daily Telegraph*

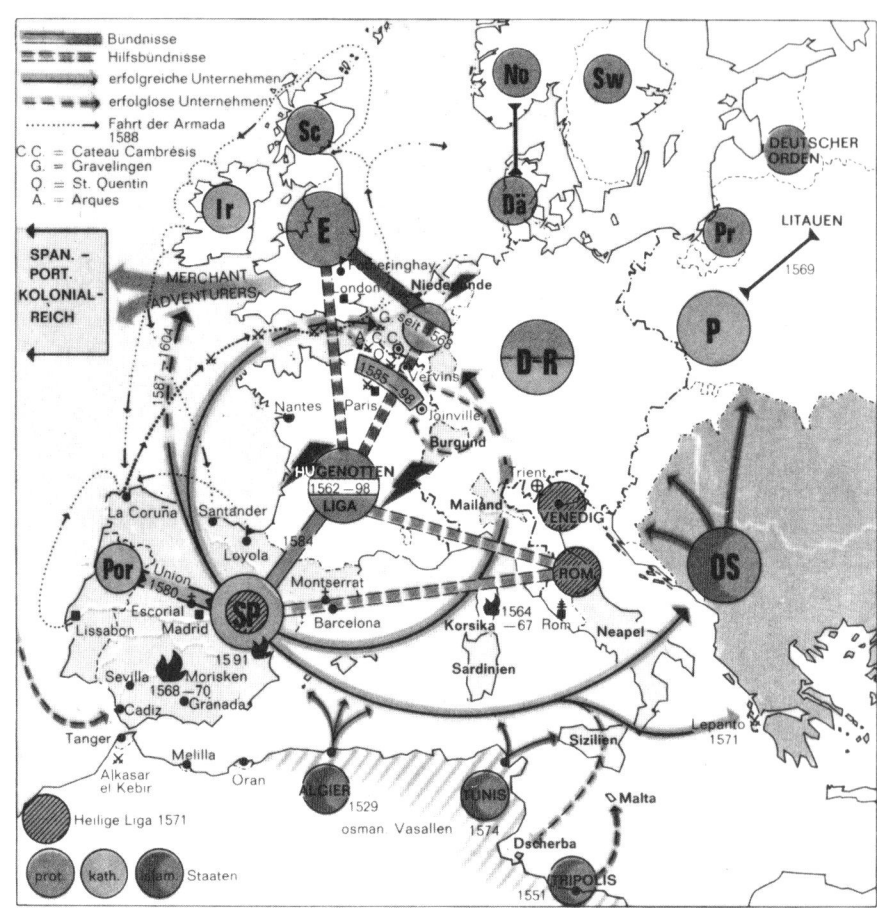

1

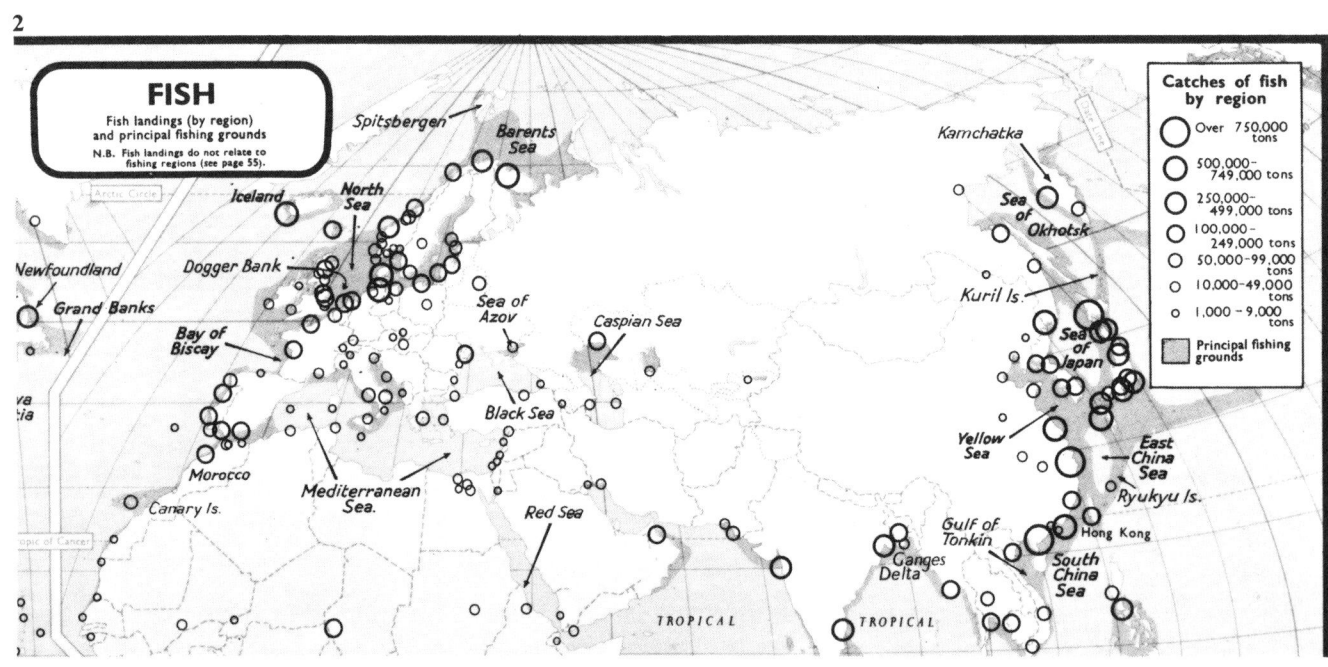

2

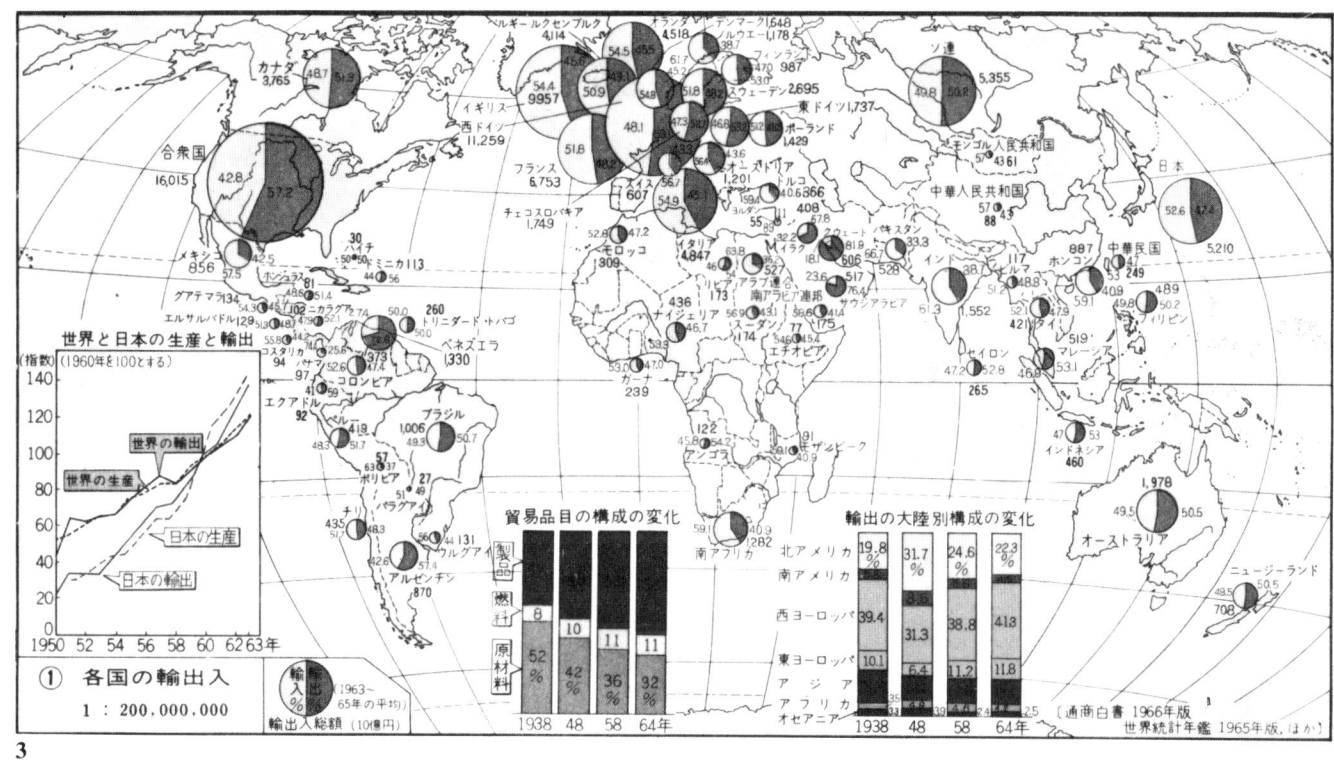

① 各国の輸出入　1:200,000,000

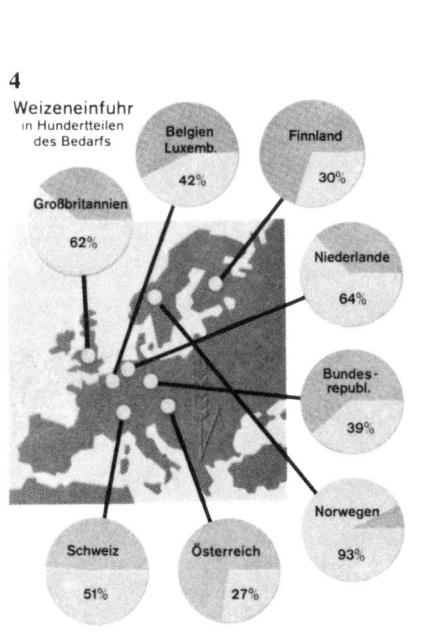

Weizeneinfuhr in Hundertteilen des Bedarfs

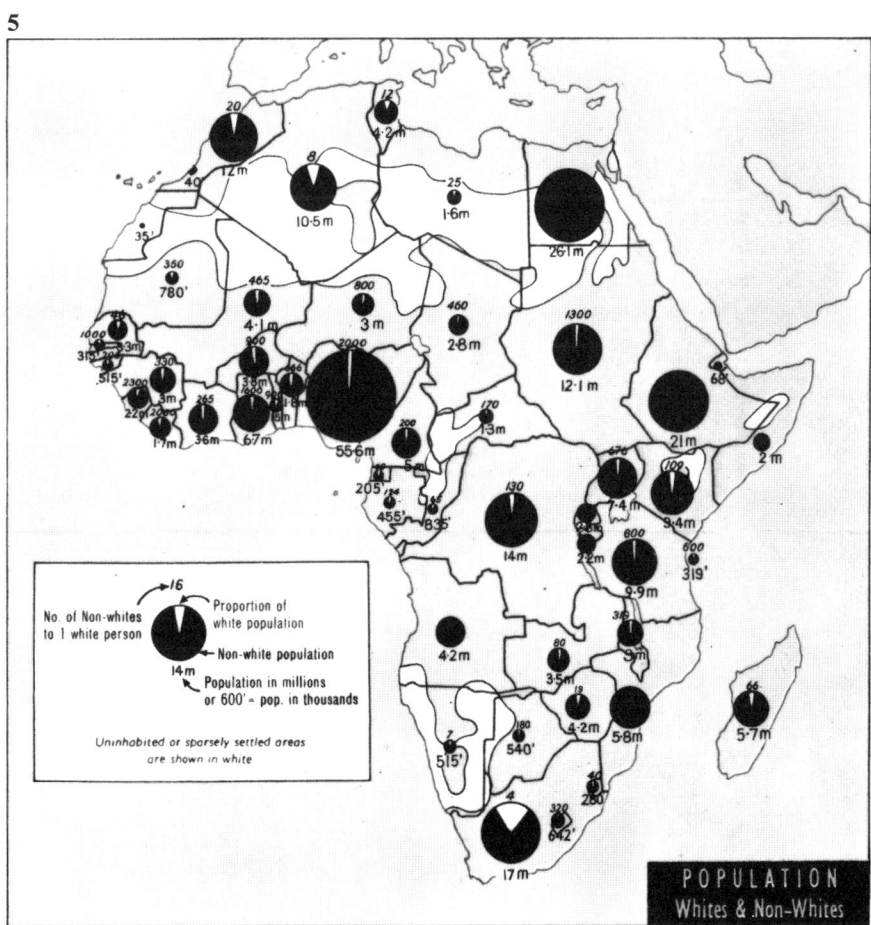

POPULATION Whites & Non-Whites

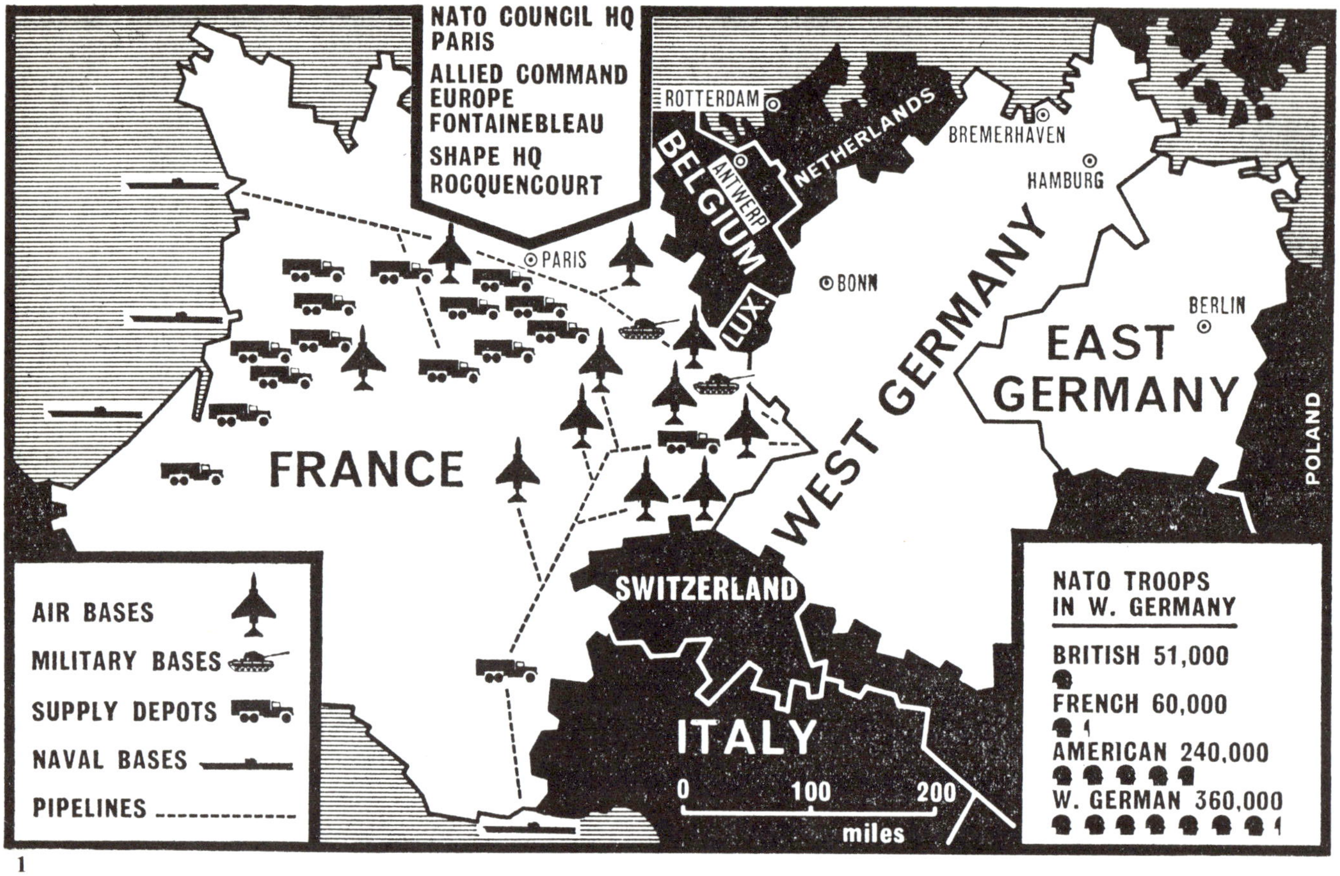

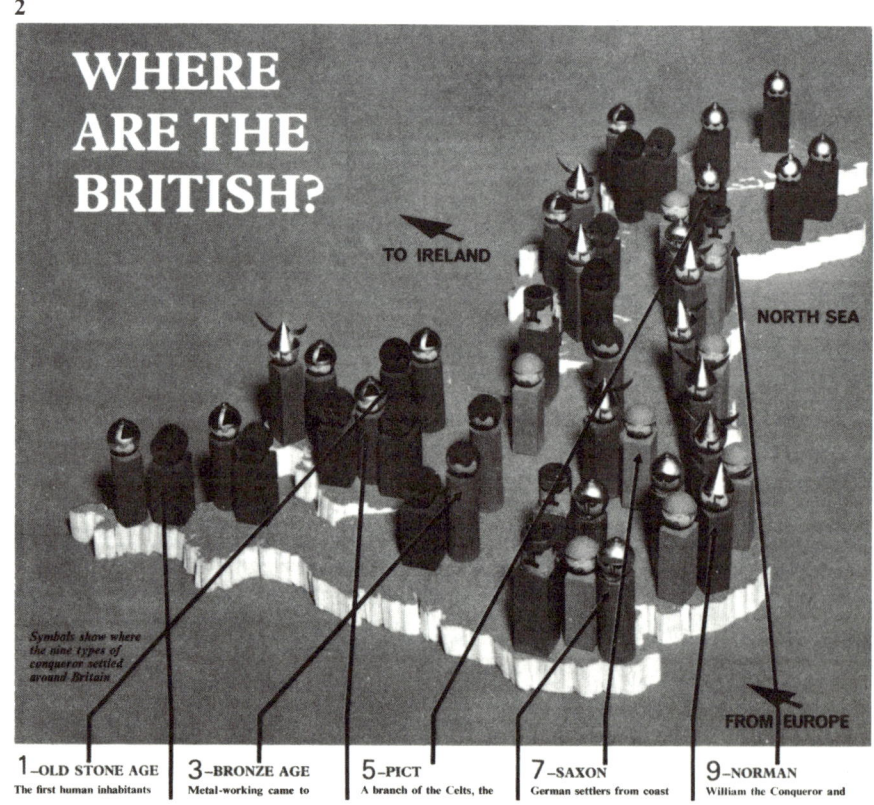

Symbols – non-quantitative

1 Map from a newspaper, showing the NATO network in France. Symbols locate bases and supply depots. *Observer*

2 Map from a magazine, showing the nine conquering peoples and where they settled in Britain. Three-dimensional symbols located. Full page, original in colour. *Observer*

3 Map from an atlas, showing location of marine life round the coast of Japan. Symbols need not always be simple and bold – they can be sensitive and subtle – there are over ten different fishes shown on this map.

4 Map from an atlas, showing population density and the religions of Asia. Symbols forming an overall texture to show extent of various religions. Explanatory. *New Relief World Atlas*, Paul Hamlyn

Symbols – quantitative

1 Map from an atlas, showing exports from Britain to Europe. The problems are identifiable symbols, large numbers of symbols obscuring the map, and adding up the symbols to get the total amount exported. Original in full colour. *Complete Atlas of the British Isles,* Reader's Digest

2 Map from a school text book, showing production of some important metals. Figures by the side of symbols avoid representation by the repetition of the basic symbols. *Chemicals and where they come from,* Longmans/Penguin Books

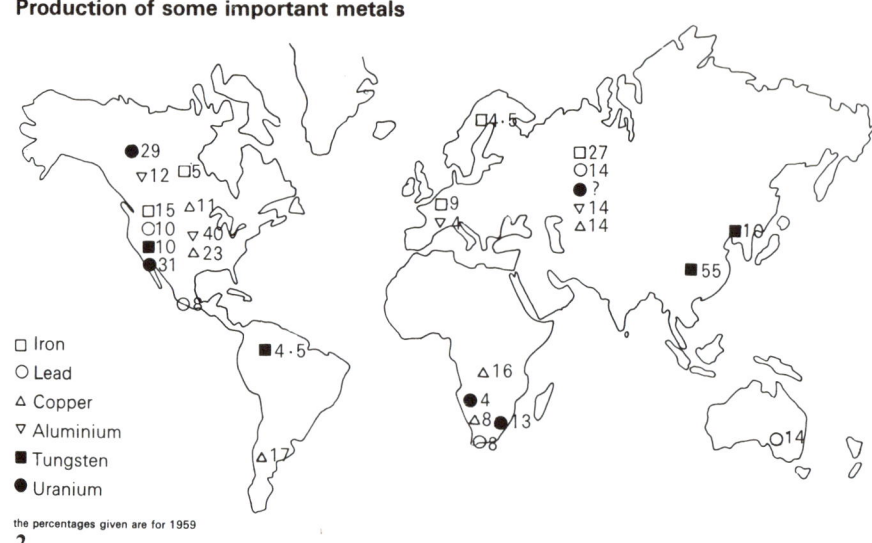

Production of some important metals

□ Iron
○ Lead
△ Copper
▽ Aluminium
■ Tungsten
● Uranium

the percentages given are for 1959

2

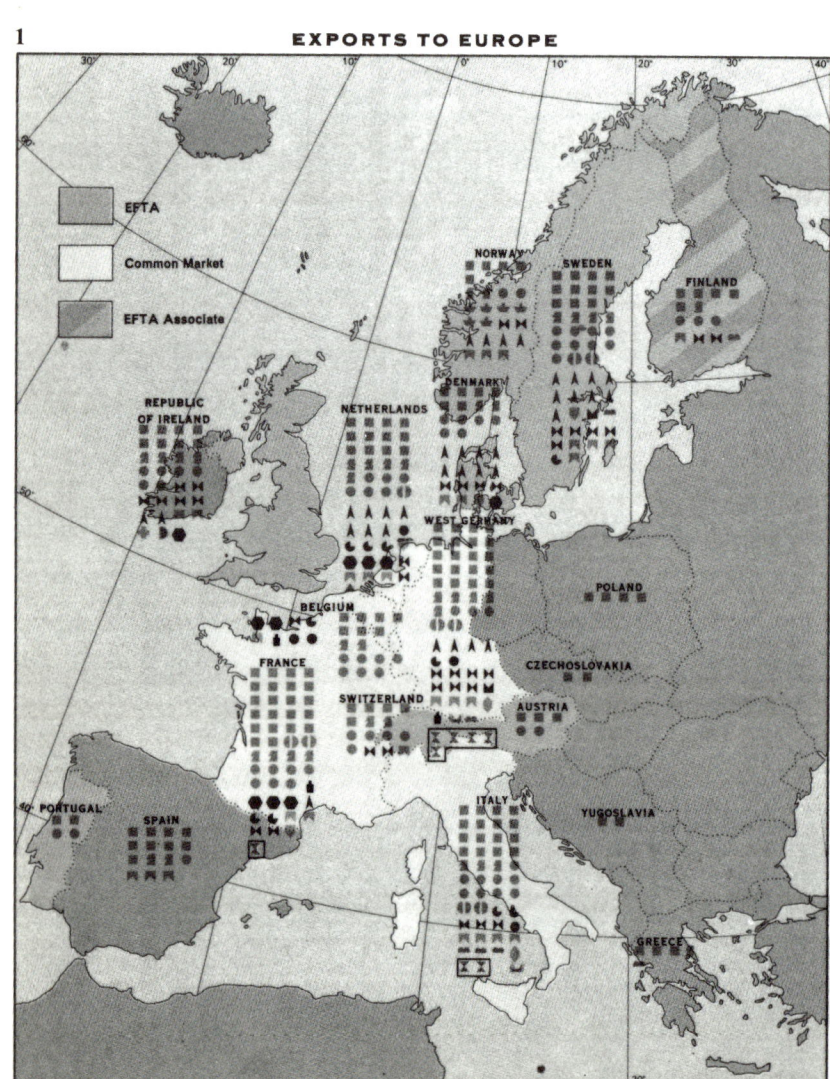

1 EXPORTS TO EUROPE

EXPORTS FROM BRITAIN
(each symbol represents £2·5 million)

RAW MATERIALS

MINERALS
- *Petroleum* and its products, including motor spirit, kerosene, distillate fuels, lubricating oils, greases
- *Chemicals:* elements, organic and inorganic compounds
- *Copper* and copper alloys
- *Coal* including anthracite, coke, briquettes

ANIMAL, VEGETABLE DERIVATIVES; SUBSTITUTES
- *Rubber:* natural, synthetic, reclaimed
- *Dyeing, tanning, colouring materials:* pigments, paints, varnishes, printing ink
- *Hides and skins* (undressed); furs

NATURAL FIBRES
- *Wool:* sheep's and lambs'

MANUFACTURED GOODS

MINERALS
- *Iron and steel:* bars, rods, wire, plates, tubes, railway lines
- *Plastics materials:* regenerated cellulose, artificial resins
- *Silver and platinum:* bullion, ingots, grains, powders
- *Pottery:* porcelain, china, earthenware
- *Medicinal and pharmaceutical:* proprietary medicines, medicaments, wadding, bandages, antibiotics, vaccines

BEVERAGES
- *Whisky:* Scotch or Irish

ANIMAL AND VEGETABLE MANUFACTURES
- *Paper and paperboard:* newsprint, tissue paper, printing paper, wallpaper, cigarette paper, hardboard, cardboard
- *Leather* from cattle, sheep, goats; dressed fur skins
- *Footwear* of leather and rubber

TEXTILES
- *Clothing:* garments, headgear, mainly of wool and cotton
- *Yarn and fabrics:* wool, linen, cotton, man-made yarns and fibres

MECHANICAL ENGINEERING
- *Machinery, other than electrical:* machinery used in agriculture, offices, textile and leather making, metal working; pumps, mechanical handling equipment, heating and cooling equipment, steam and internal combustion engines
- *Electrical equipment:* power machinery and switch gear, generators, electric motors, cable and wire, telecommunications equipment, domestic electrical equipment
- *Motor vehicles:* cars, motor-cycles, lorries, vans, ambulances, spare parts, accessories
- *Aircraft:* aeroplanes, gliders
- *Railway vehicles:* locomotives, rolling stock
- *Ships and boats*
- *Bicycles*
- *Instruments:* professional, scientific, process controlling; photographic and optical goods, watches, clocks

□ Re-exports

3 Map from a magazine, showing the Gaullist (U.D.R. and R.I.) success in the French elections. Compare this map with page 91. *Paris Match*

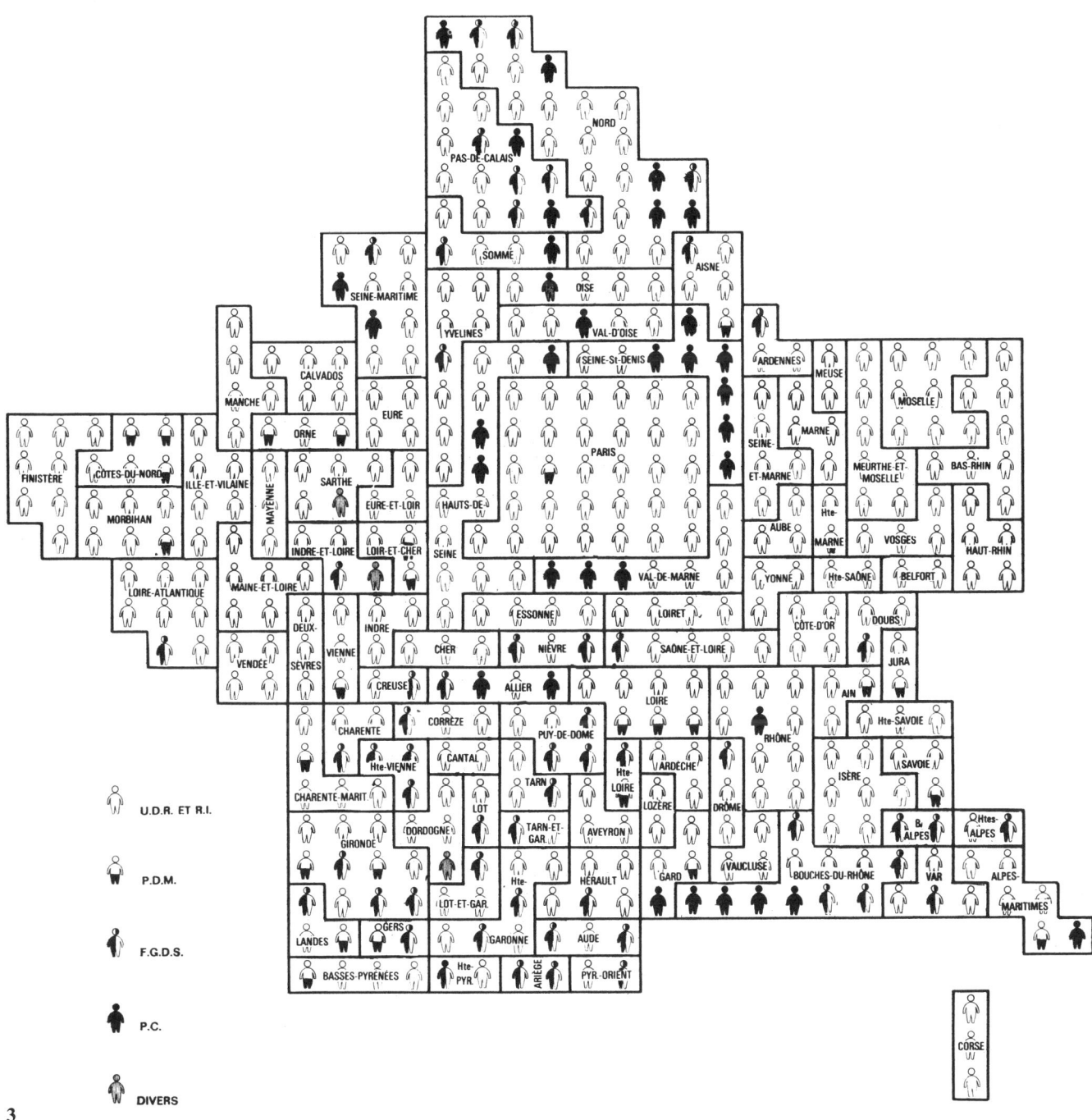

Distorted map (cartogram)

1   Map from a book, showing main administrative regions of the Soviet Union in 1944. Comparison of areas on a conventional map is not always easy, but by 'building' a map from an identical unit such as the square, evaluation of various areas is made easier. An Isotype map. K. E. Holme, *Two Commonwealths,* Harrap

2   Map from a magazine, showing high and low air pressure over United States. *Power*

3   Map from a book, showing the area of the bases of various countries in proportion to their population – the height representing their per capita income. Isometric perspective. Erwin Raisz, *Principles of Cartography,* McGraw-Hill

4   Map from a United Nations publication, showing the size of the country in proportion to its population, shaded according to the daily intake of calories. UNESCO *Courier*

5   Map from a business magazine, showing the distribution of top level United States scientists. The size of each state is proportional to its population of scientists. It is colour coded to show gain or loss of scientists. *Fortune*

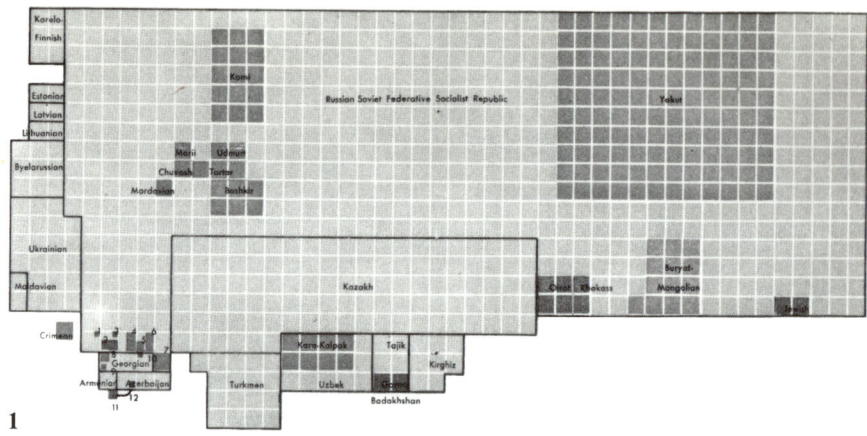

1

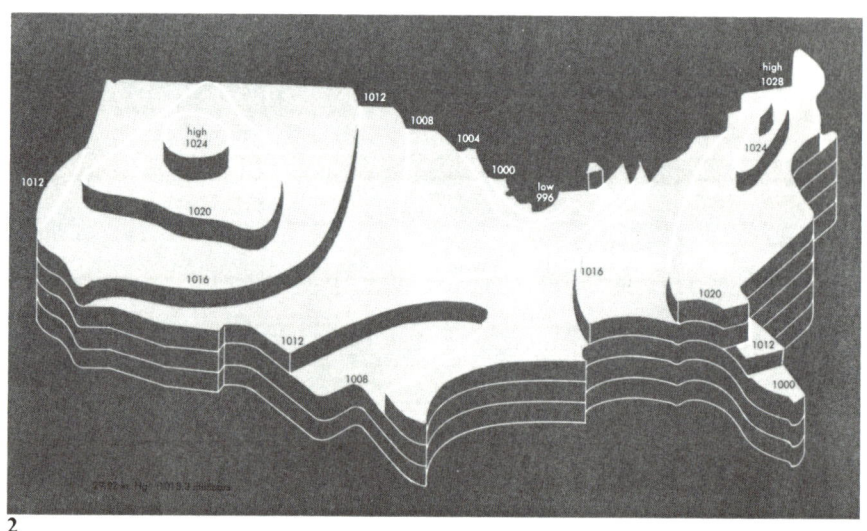

2

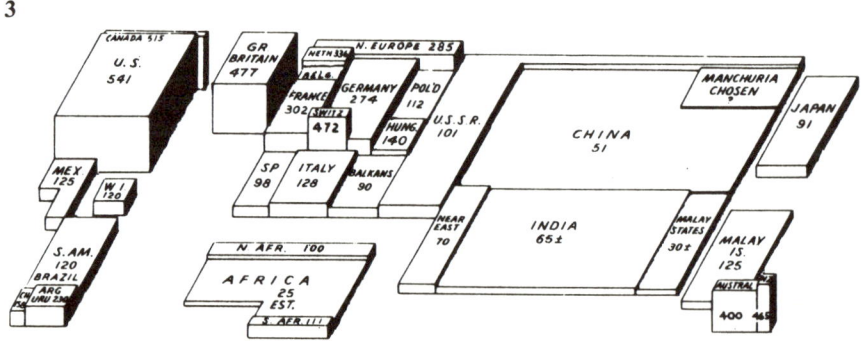

3

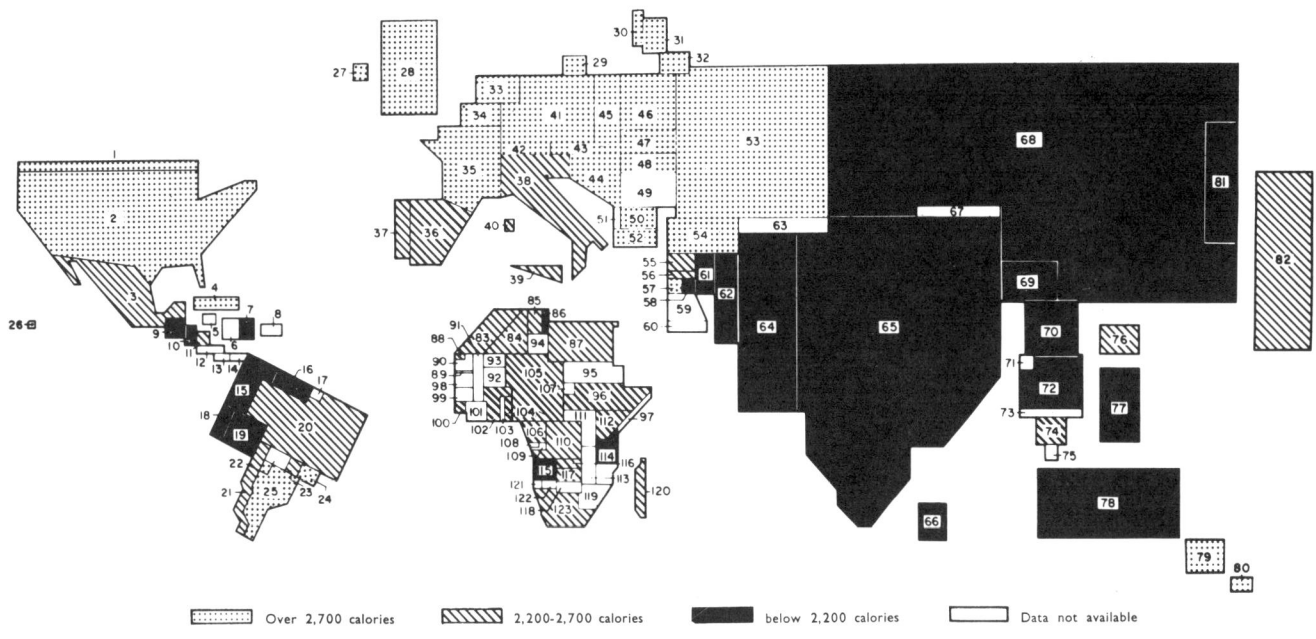

▦ Over 2,700 calories	▨ 2,200-2,700 calories	■ below 2,200 calories	☐ Data not available	

1. Canada	22. Bolivia	43. Austria	64. Pakistan	85. Tunisia	106. Cameroun
2. United States	23. Paraguay	44. Yugoslavia	65. India	86. Libya	107. Central African
3. Mexico	24. Uruguay	45. Eastern Germany	66. Ceylon	87. United Arab Republic	Republic
4. Cuba	25. Argentina	46. Poland	67. Nepal	88. Mauritania	108. Gabon
5. Jamaica	26. Hawaii	47. Czechoslovakia	68. China, Mainland	89. Gambia	109. Congo - Brazzaville
6. Haiti	27. Ireland	48. Hungary	69. Burma	90. Senegal	110. Congo - Leopoldville
7. Dominican Republic	28. United Kingdom	49. Rumania	70. Thailand	91. Mali	111. Uganda
8. Puerto Rico	29. Denmark	50. Bulgaria	71. Laos	92. Upper Volta	112. Kenya
9. Guatemala	30. Norway	51. Albania	72. Viet-Nam & N. Viet-Nam	93. Niger	113. Ruanda-Urundi

4

5

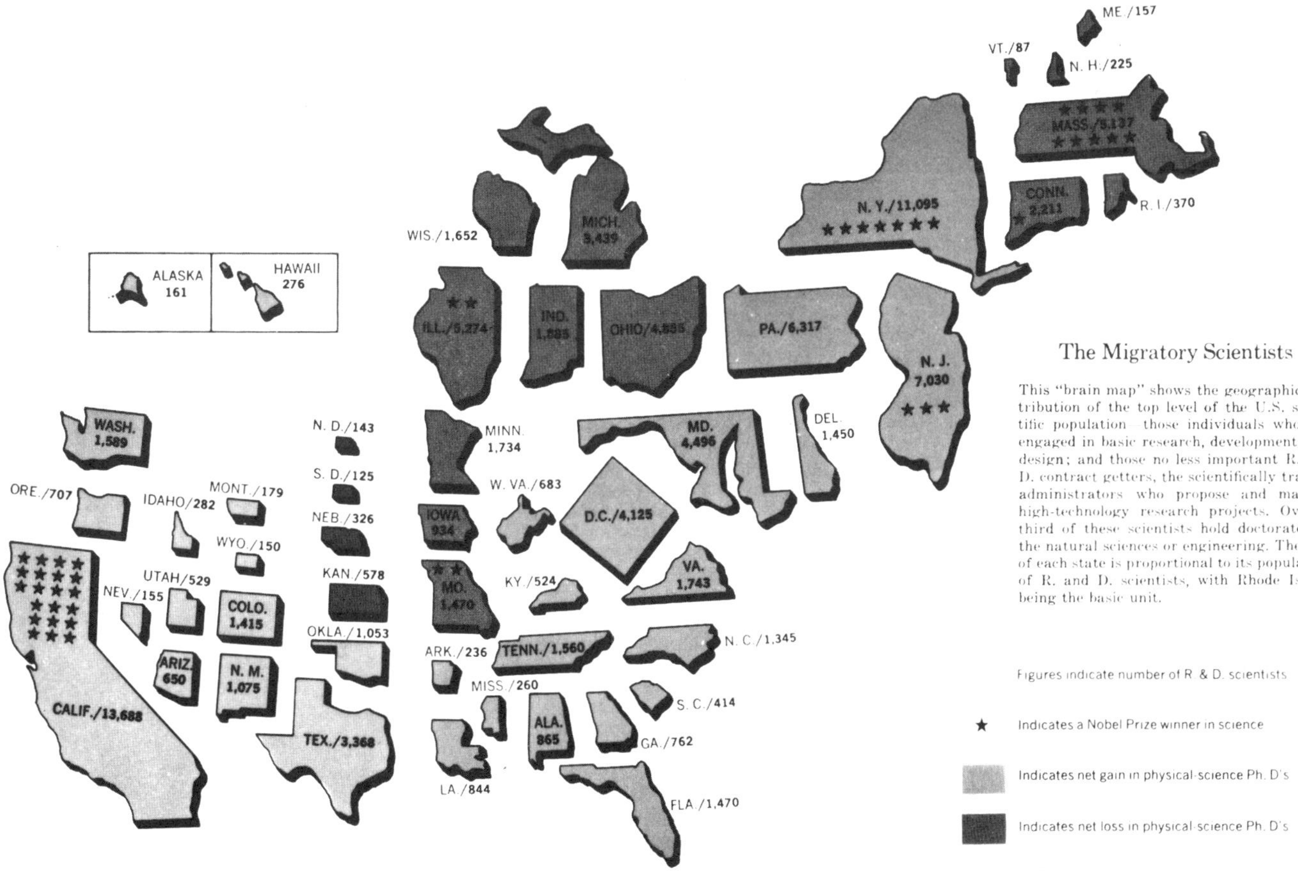

The Migratory Scientists

This "brain map" shows the geographic distribution of the top level of the U.S. scientific population—those individuals who are engaged in basic research, development, and design; and those no less important R. and D. contract getters, the scientifically trained administrators who propose and manage high-technology research projects. Over a third of these scientists hold doctorates in the natural sciences or engineering. The size of each state is proportional to its population of R. and D. scientists, with Rhode Island being the basic unit.

Figures indicate number of R. & D. scientists

★ Indicates a Nobel Prize winner in science

▨ Indicates net gain in physical-science Ph. D's

■ Indicates net loss in physical-science Ph. D's

Distorted map (cartogram)

1 Map from a magazine, locating places according to their time-distance from London by the fastest moving service to the town. All the major cities of Britain appear almost equidistant from London. But apart from the main cities, certain districts appear very remote. Distorting a map by time travelled. *New Society*

2 Map from an atlas, showing the counties of Britain divided into electoral constituencies. The number of electors in constituencies is broadly equal and the map makes the area of each constituency the same. The map directly indicates distribution of electors, and since the proportion of electors to population is fairly constant, the map also shows the distribution of the population. Distortion according to the electoral strength. The map was constructed from 40,320 small squares. *Dr T. H. Hollingsworth*

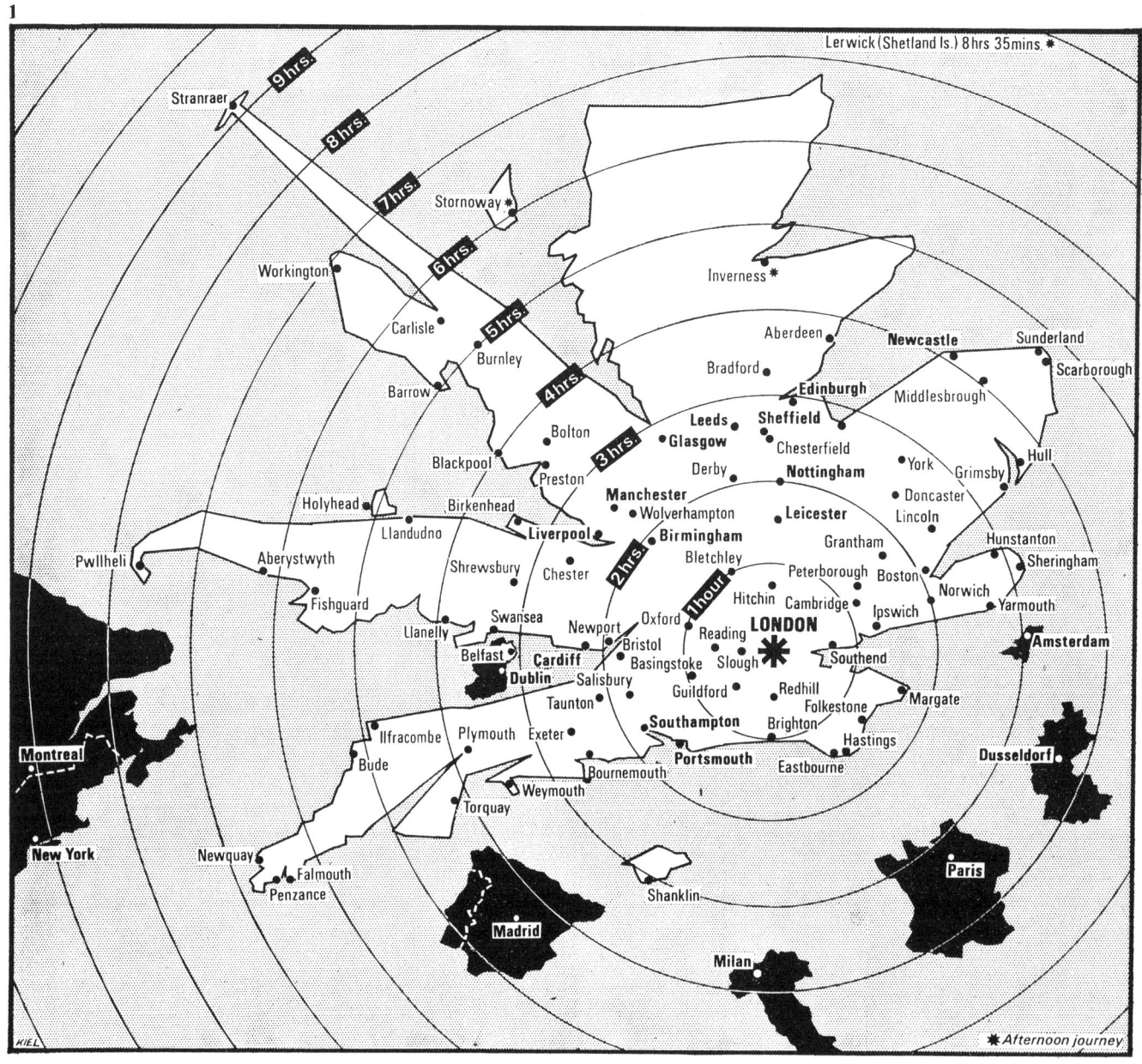

WHERE THE VOTERS ARE
THE SHAPE OF ELECTORAL STRENGTH IN THE COUNTIES

THE POPULATION of the United Kingdom is divided into constituencies, each of which elects one Member of Parliament. To take account of the movement of population, the House of Commons (Redistribution of Seats) Act 1949, requires constituency boundaries to be re-drawn at least once every 15 years; the last revision was in 1954 and a further review was announced in 1965. There is a Boundary Commission for each country, whose first task is to determine the number of seats for England, Scotland, Wales and Monmouthshire, and Northern Ireland. Great Britain must have not substantially more or less than 613 seats; Scotland, not less than 71; Wales, not less than 35; and Northern Ireland must have 12.

The principle of the division of the population—that the number of electors in constituencies in any one country should be broadly equal—is reflected by this demographic map in which the area of each constituency is the same. (The map was constructed from 40,320 small squares, and while geographic features are only roughly preserved each constituency is bounded by its true neighbours.) However, constituencies in Scotland and Wales have a slightly smaller number of electors, and in Northern Ireland a slightly larger number, so that Scotland and Wales appear larger, and Northern Ireland smaller than on a true demographic map.

The map directly indicates the distribution of electors; since the proportion of electors to the rest of the population is fairly constant in different areas, the map also reflects the distribution of population.

GROWTH OF THE POPULATION AND THE ELECTORATE

In 1832 the franchise was extended to the urban middle class, in 1867 to most urban workers, in 1884 to most rural workers, in 1918 to men over 21 and women over 30, and in 1928 to women over 21.

ENGLAND		WALES	
1	Greater London	41	Glamorgan
2	Kent	42	Carmarthen
3	Sussex	43	Pembroke
4	Surrey	44	Cardigan
5	Hampshire & Isle of Wight	45	Brecknock
6	Dorset	46	Radnor
7	Devon	47	Montgomery
8	Cornwall	48	Merioneth
9	Somerset	49	Denbigh
10	Gloucester	50	Flint
11	Wiltshire	51	Caernarvon
12	Berkshire	52	Anglesey
13	Oxford		
14	Buckingham	**SCOTLAND**	
15	Worcester	53	Dumfries
16	Hereford	54	Kirkcudbright
17	Monmouth	55	Wigtown
18	Shropshire	56	Ayr
19	Stafford	57	Bute
20	Warwick	58	Renfrew
21	Cheshire	59	Lanark
22	Lancashire	60	Argyll
23	Westmorland	61	Dunbarton
24	Cumberland	62	Inverness
25	Northumberland	63	Ross & Cromarty
26	Durham	64	Sutherland
27	Yorkshire	65	Caithness
28	Derby	66	Orkney Islands
29	Nottingham	67	Shetland Islands
30	Lincoln	68	Nairn
31	Leicester	69	Moray
32	Northampton	70	Banff
33	Rutland	71	Aberdeen
34	Bedford	72	Kincardine
35	Hertford	73	Angus
36	Huntingdon & Peterborough	74	Perth
37	Cambridge & Isle of Ely	75	Stirling
38	Norfolk	76	Kinross
39	Suffolk	77	Clackmannan
40	Essex	78	Fife
		79	West Lothian
		80	Midlothian
		81	Peebles
		82	Selkirk
		83	Roxburgh
		84	East Lothian
		85	Berwick
		NORTHERN IRELAND	
		86	Antrim
		87	Down
		88	Armagh
		89	Tyrone
		90	Fermanagh
		91	Londonderry

Ray map

1 Map from a school textbook, showing Royal Flying Doctor service network in 1962. Many centres and their range shown on one map. G. R. Davidson, *Medicine Through the Ages*, Methuen

2 Part of a wall map of Africa, showing 'crow fly' distances from African towns to five other cities. Lines cut short and distance in miles added. *Map of Africa*, Daily Telegraph

3 Map from a magazine, showing exact positions of ships and planes over the Atlantic at 1600 hours, 21 July 1965. Original in colour. *Observer*

4 Diagram from an advertisement, showing airline routes to cities in Europe. Only the 'ray' feeling survives, as actual distance and exact geographical location are distorted. Swissair

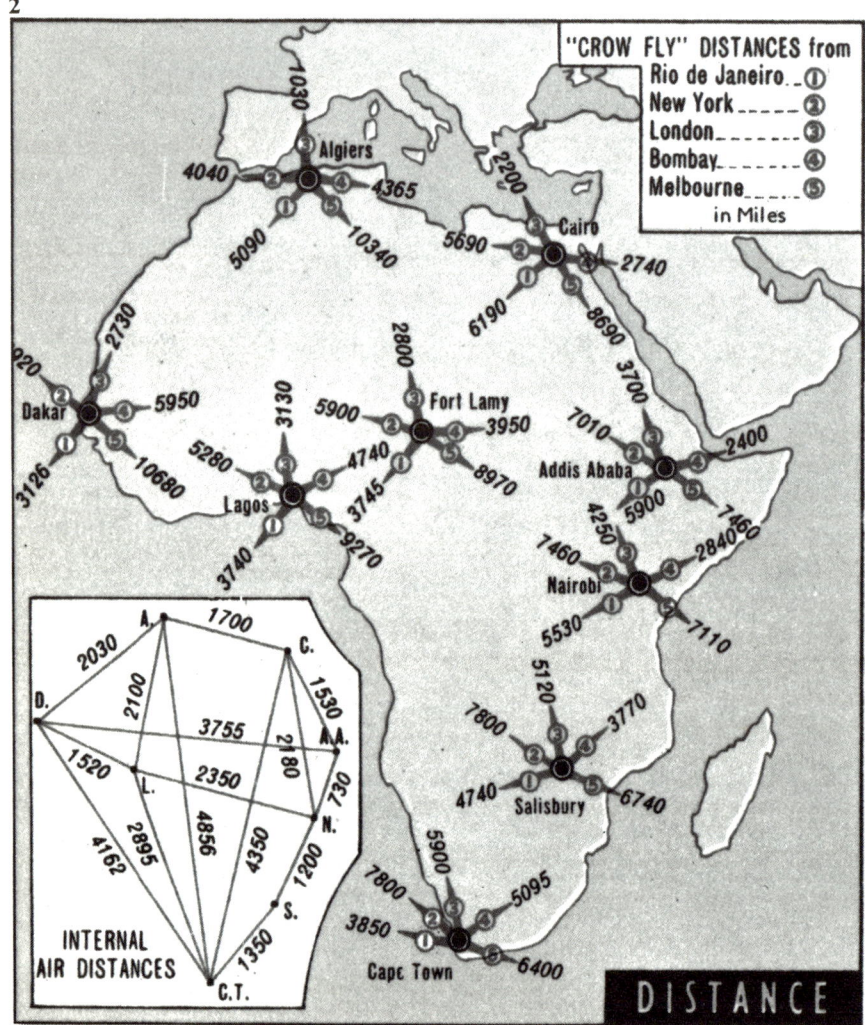

THE TRANSATLANTIC TRAFFIC JAM
1600 HOURS: 21 JULY 1965

This summer will be the busiest transatlantic travel time yet. Here, on an afternoon in July arbitrarily chosen for heavy traffic (at that time of day most planes are heading westbound) 56 planes and 13 ships are on, or over, the Atlantic. They are:

Ships
1 Rafaello
2 Michelangelo
3 Independence
4 Constitution
5 Shalom
6 United States
7 Olympia
8 France
9 Empress of England
10 Rotterdam
11 Nieuw Amsterdam
12 Queen Mary
13 Queen Elizabeth

Airlines
A El-Al
B Qantas
C Air France
D Air India
E TWA
F Air Lingus
G Swissair
H Pan American
K BOAC
L KLM
M SAS
N Lufthansa
O CPA
P Alitalia
Q Sabena
R Air Canada
S Iberia

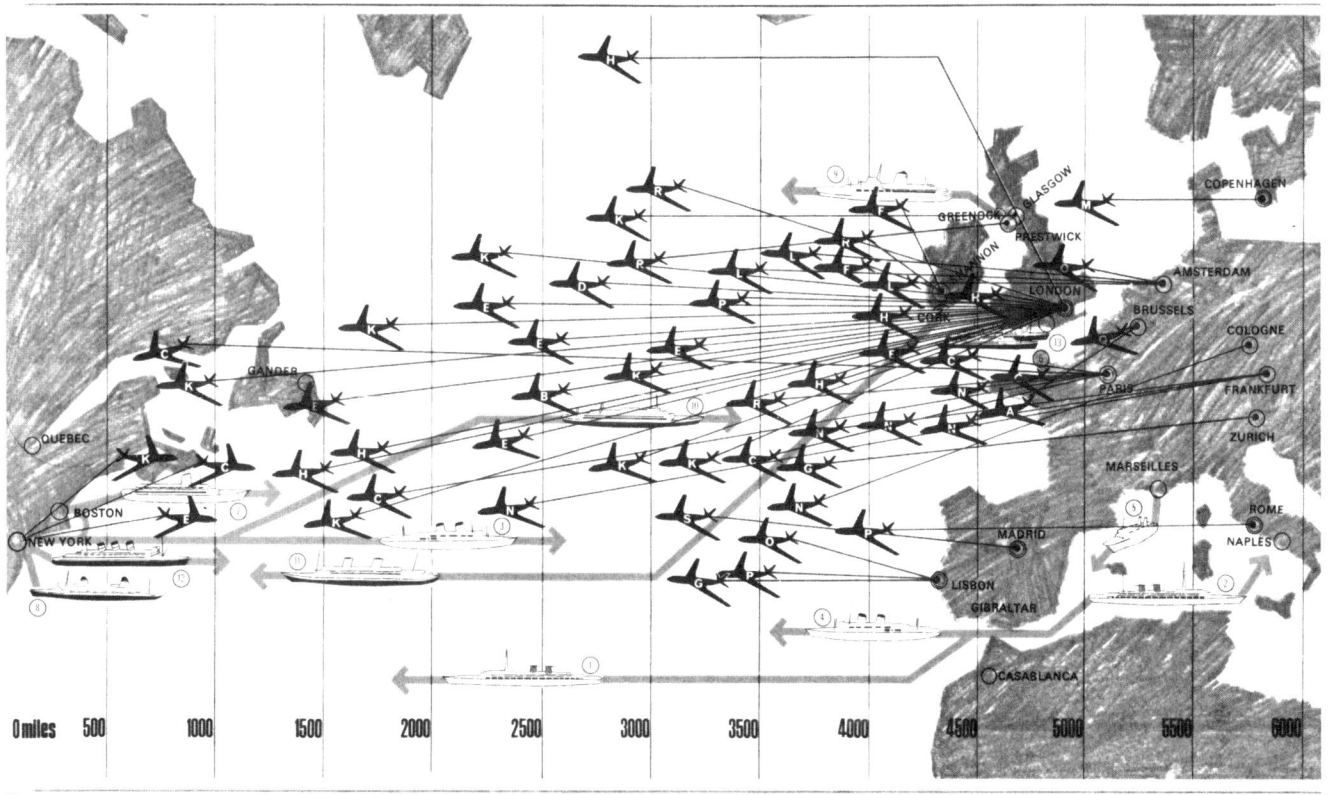

3

4

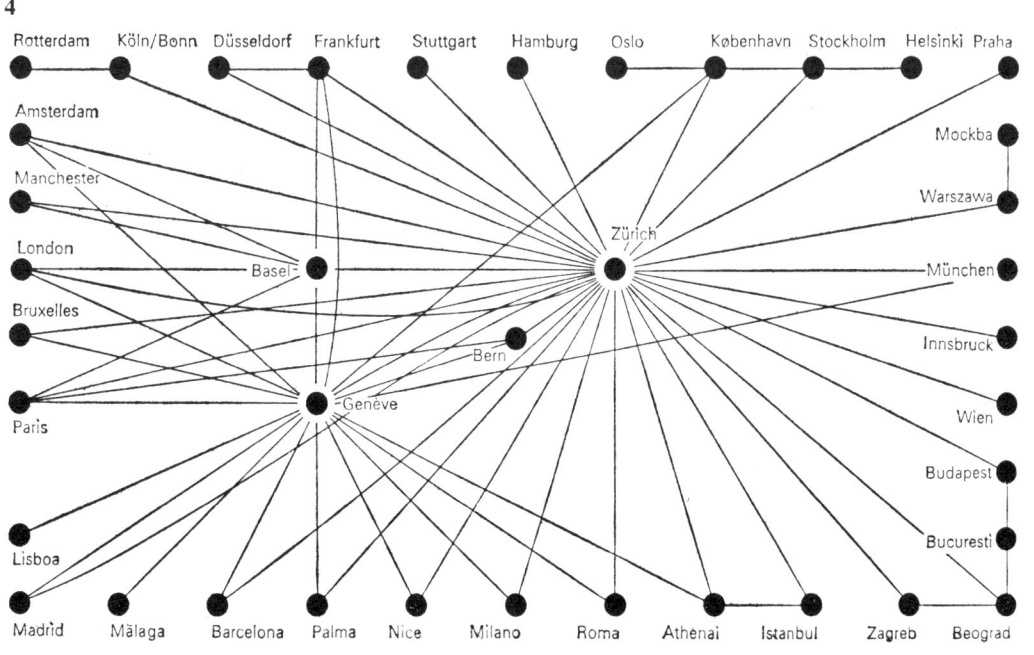

Flow line map

1 Map from book, showing the present-day peak-hour traffic flow for Newbury, Berks, England. Original has local town traffic (red) and through traffic (green). *Traffic in Towns,* HMSO

2 Map from an educational book, showing the countries that export their surplus wheat and those countries which import it. The flow of international trade in wheat is shown clearly. H. L. Edlin, *Man and Plants,* Aldus Books

3 Map from a business magazine, showing where the daily tonnage of freight of two railways will flow. Background map is omitted so that information can be clearly seen. *Fortune*

4 Map from a business magazine, showing the size of direct US investment in foreign countries at the end of 1966. Size of countries distorted in proportion to amount received. Original in colour. *Fortune*

1

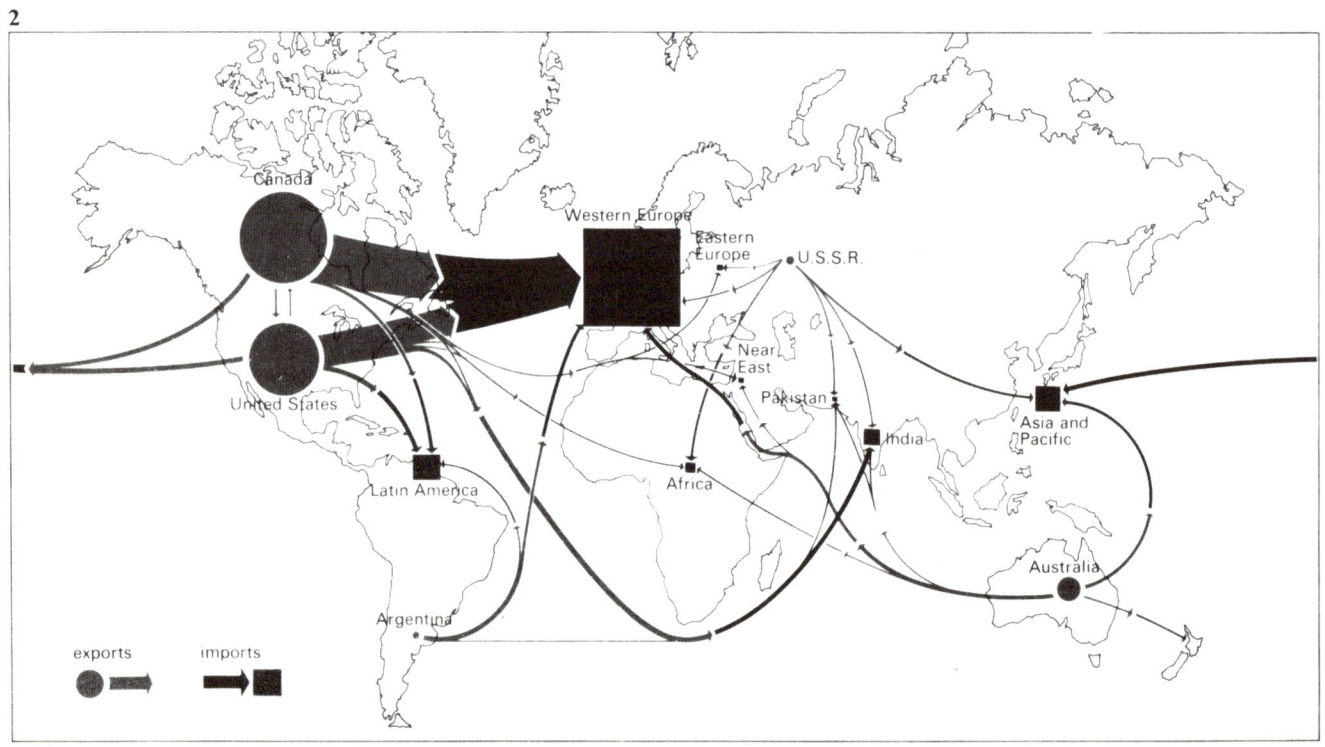

2

3

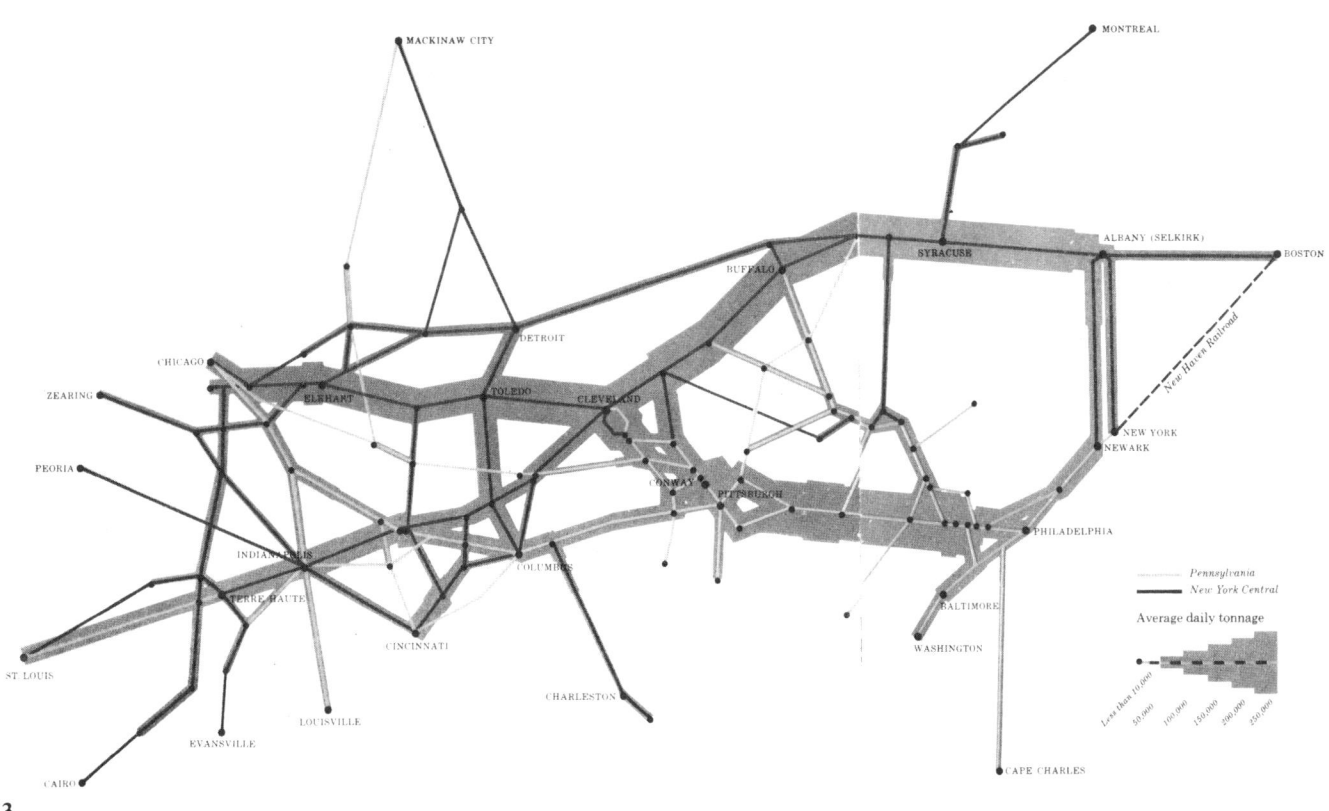

4

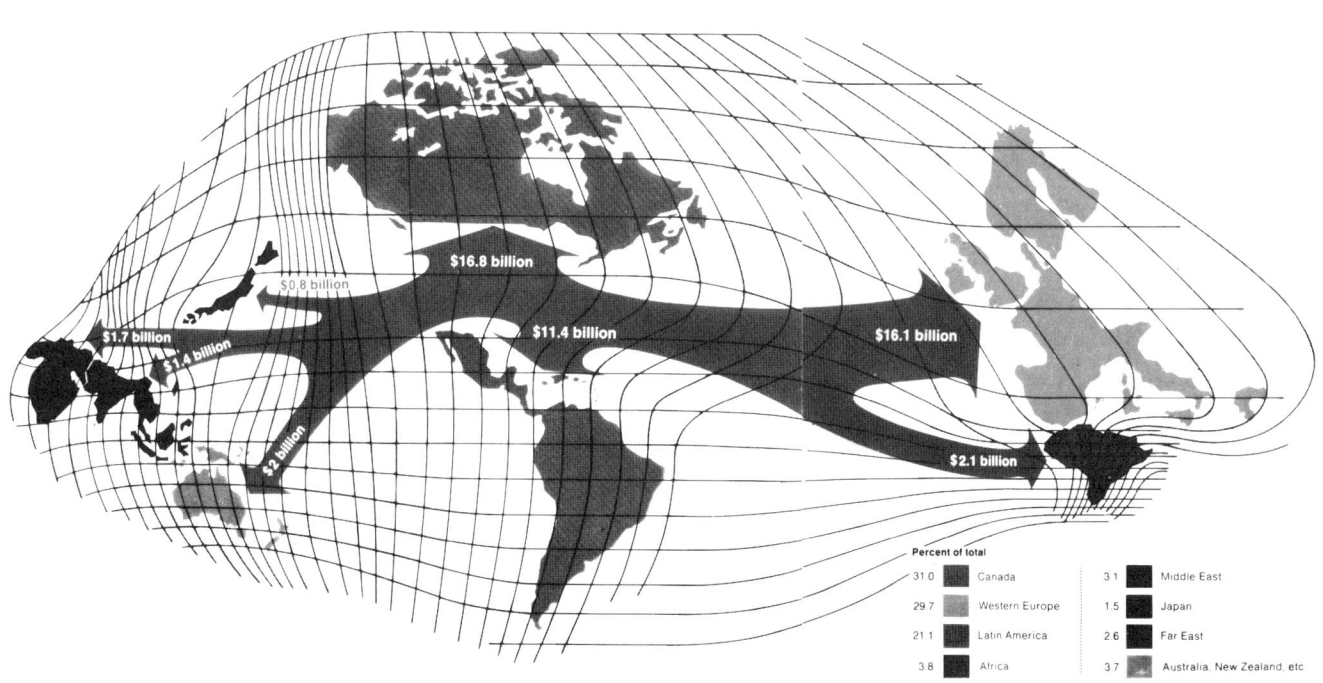

105

Flow line map becomes flow diagram

1 Map from an educational book, showing flow of crude oil and refined petroleum products. Squares compare consumption of oil in seven regions of the world. E. G. Sterland, *Energy into Power*, Aldus Books

2 Map from a text book, showing oil produced and imported for various parts of the world. Marie Neurath, *Living with One Another*, Parrish

3 Map from a scientific magazine, showing major world fuel shipments for 1960. Circles represent exporting regions, rectangles importing regions. They are proportional in size to the weight of fuel shipped or received. Circles and rectangles are divided to represent solid fuel, liquid fuels and crude oil. Width and colour of flow line indicates weight and nature of shipment. Original in two colours.
Scientific American

4 Diagram from a text book, showing oil production (black), refinery capacity, oil consumption (grey). Original in full colour. The map has been omitted – this is a flow diagram.
A. Jentzsch and J. Winkler, *Güterkunde*, Westermann

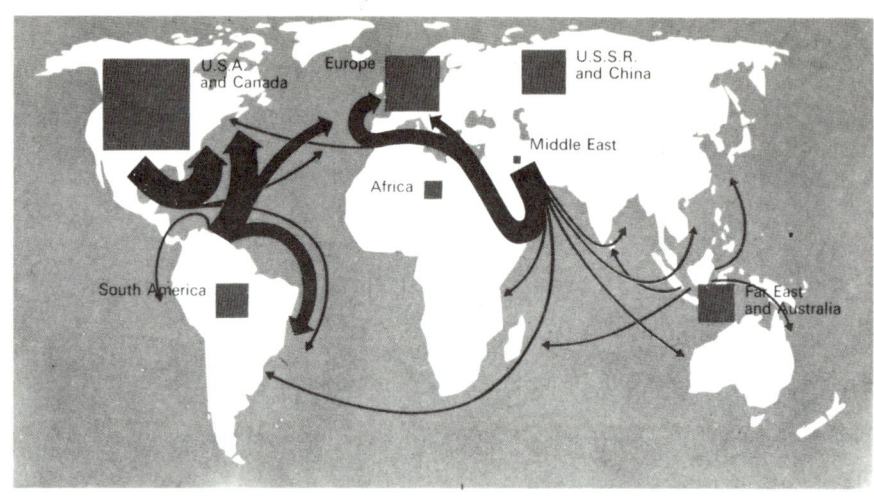

1

2
Crude Oil Production and Shipments, 1960

Each symbol represents 25 million tons of hard coal equivalent
black: home produced, on ship: exported
outline on ship: imported

106

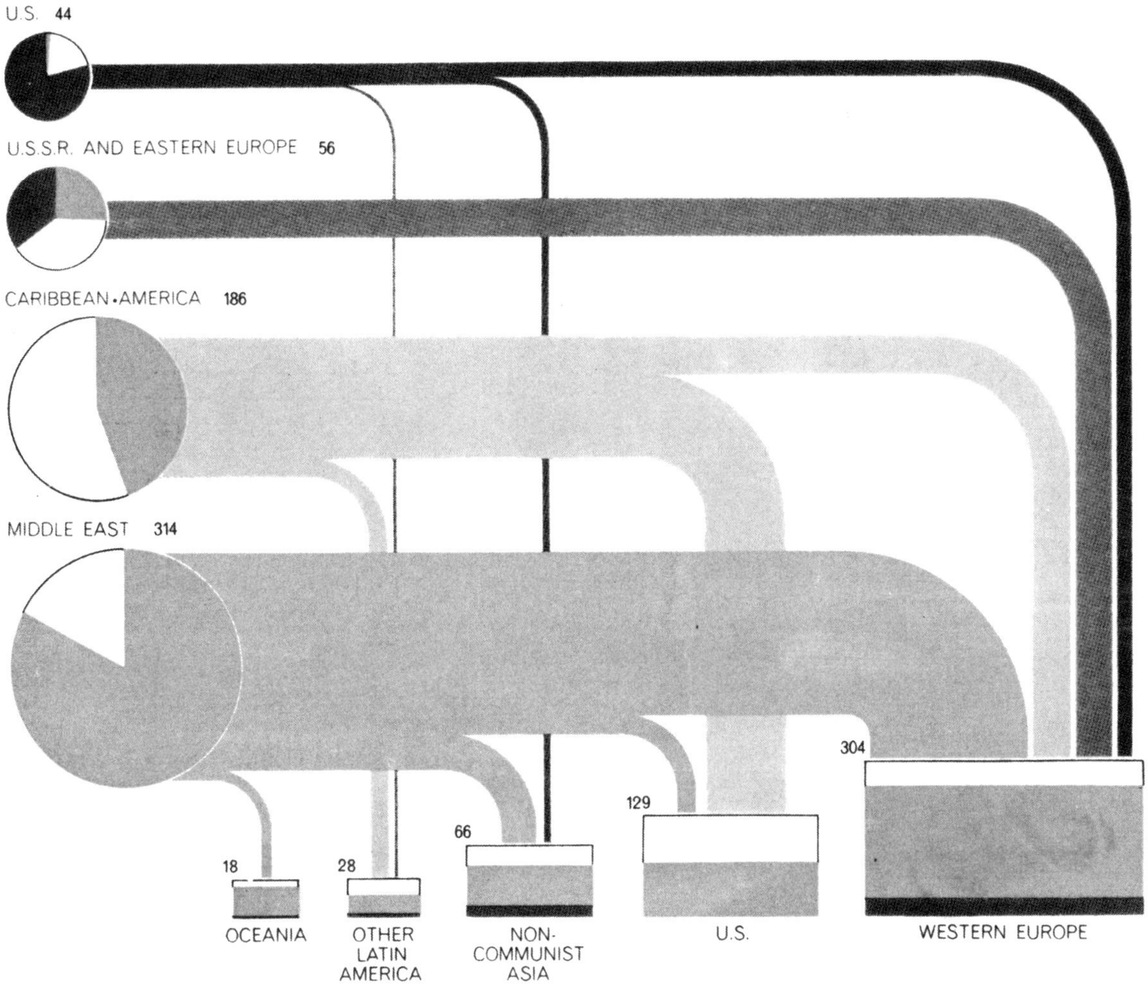

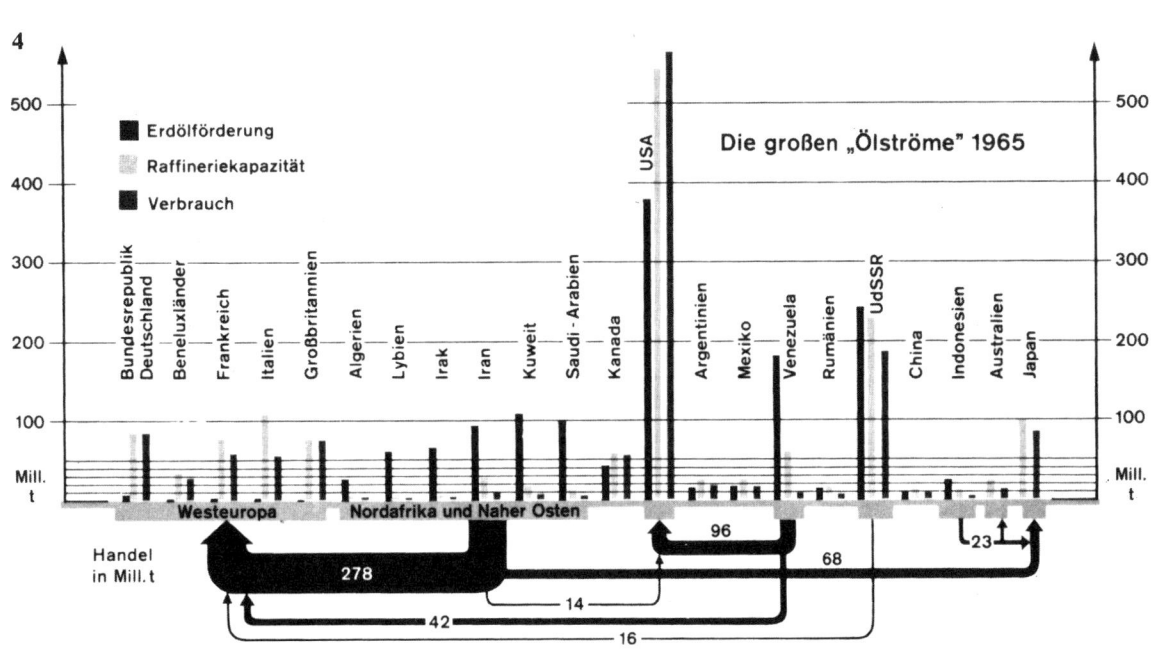

3 Explanatory diagrams

This section discusses diagrams which explain: from stages in a manufacturing process, or the structure of an organization, to events related to each other in time. These diagrams do not usually make quantitative statements, though some may be adapted to do so.

The main design problem of explanatory diagrams is reducing information to the essential without distortion. This can be done in many ways from a representation approach to a near abstract one. Various factors will influence the choice, such as the audience, limitations of size, amount of information to be conveyed, and the sources of references obtainable for the diagram.

1 Diagram from a consumer magazine, showing how the magazine tests goods for one of its reports. Original in two colours. *Which?*

2 Diagram from a business publication, showing the manufacture of rayon. Each stage clearly shown and captioned. Societé de la Viscose Suisse, Emmenbrücke

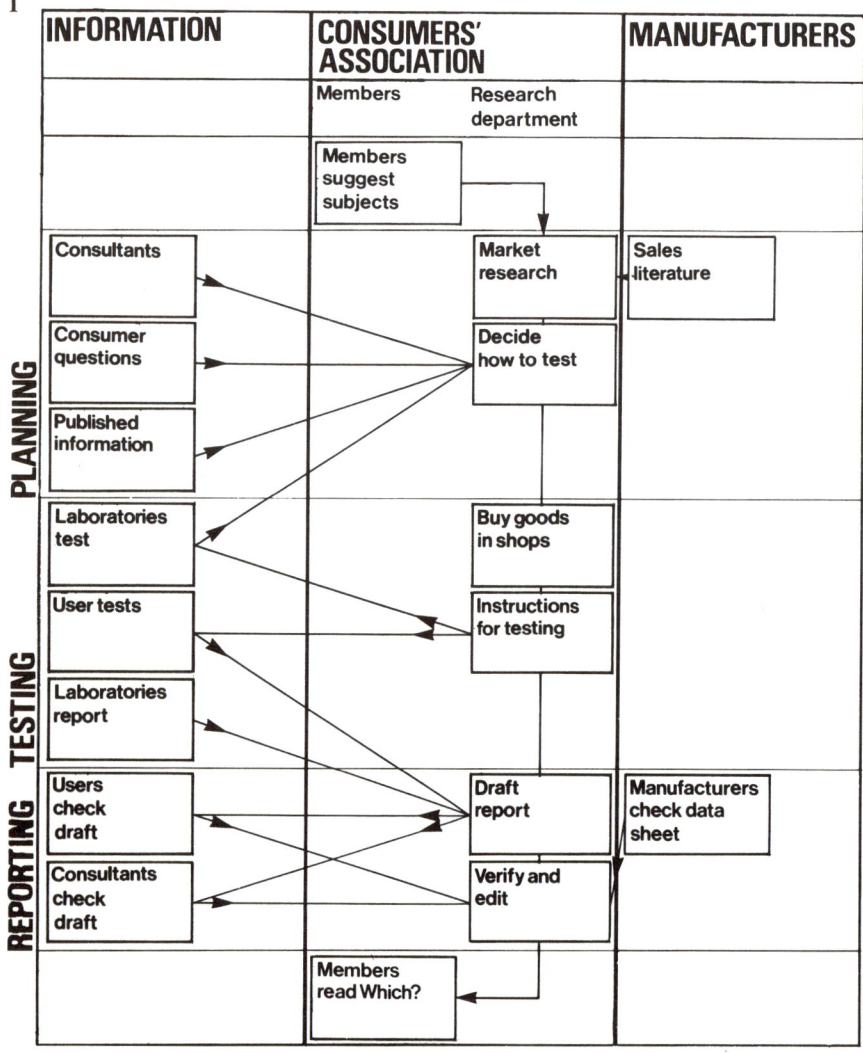

108

**Schéma de fabrication
de la rayonne et de la
FIBRANNE de VISCOSE**

RAYONNE de VISCOSE
1. Produit de base: cellulose obtenue à partir du bois
2. Préparation de l'alcali-cellulose par trempage des feuilles de cellulose dans la soude caustique
3. Déchiquetage des feuilles pressées d'alcali-cellulose
4. Mûrissement de l'acali-cellulose en cave
5. Transformation de l'alcali-cellulose en xanthate de cellulose par action du sulfure de carbone
6. Dissolution du xanthate de cellulose dans la soude caustique. Formation de la viscose
7. Filtrage de la viscose
8. Mûrissement de la viscose
9. Dégazage sous vide de la viscose
10. Filature de la rayonne par coagulation de la viscose en bain acide
11. Retordage en un seul fil des brins sortis de la filière et formation des gâteaux dans le pot-turbine
12. Désacidification, désulfuration et ensimage des gâteaux de filature
13. Séchage et conditionnement des gâteaux de filature
14. Bobinage pour la transformation ultérieure

FIBRANNE de VISCOSE
15. Filature de la FIBRANNE (analogue à celle de la RAYONNE) sous forme de grosses mèches continues
16. Désacidification, désulfuration et ensimage de la mèche continue
17. Coupage de la mèche en fibres discontinues de longueurs désirées. Formation de la FIBRANNE
18. Séchage de la FIBRANNE
19. Présentation finale de la FIBRANNE, type coton ou laine, pour la filature

PONTESA® = marque déposée par la VISCOSE SUISSE, Emmenbrücke, pour tout produit contrôlé, fabriqué avec sa viscose et répondant à ses standards.

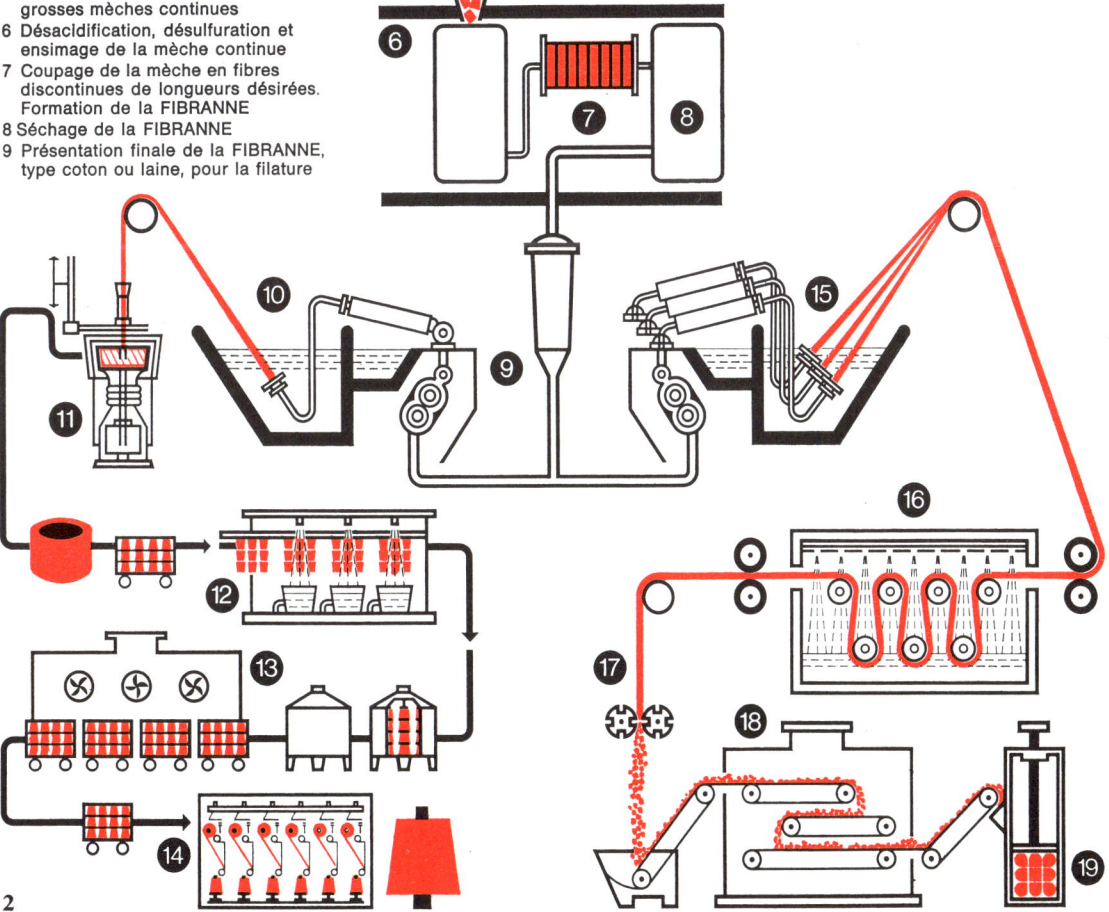

Flow chart

This diagram shows successive movements through a process from the starting point to the finish. The simplest kind of flow diagram is typographical and resembles the organization diagram (page 118). It has arrows indicating direction of flow. A more complex one is the diagram which shows a manufacturing process, depicting each stage and the machinery involved.

The lines or arrows can be coded to show the differing nature of the flow at various stages. The flow chart is normally non-quantitative, but it can show quantities by the varying thickness of flow line, as in the flow line map.

1 Diagram from an educational book, showing energy flow for a car travelling at 30 mph along a level road. Most energy is lost in cooling system and exhaust gases. The width of line shows this clearly. E. G. Sterland, *Energy into Power*, Aldus Books

2 Diagram from a paperback encyclopedia, showing annual energy balance for Switzerland. Energy from various sources, used in various ways. Vast energy loss apparent by thickness of line. Quantitative flow chart. *Das Fischer Lexikon, Technik*, Fischer Bücherei

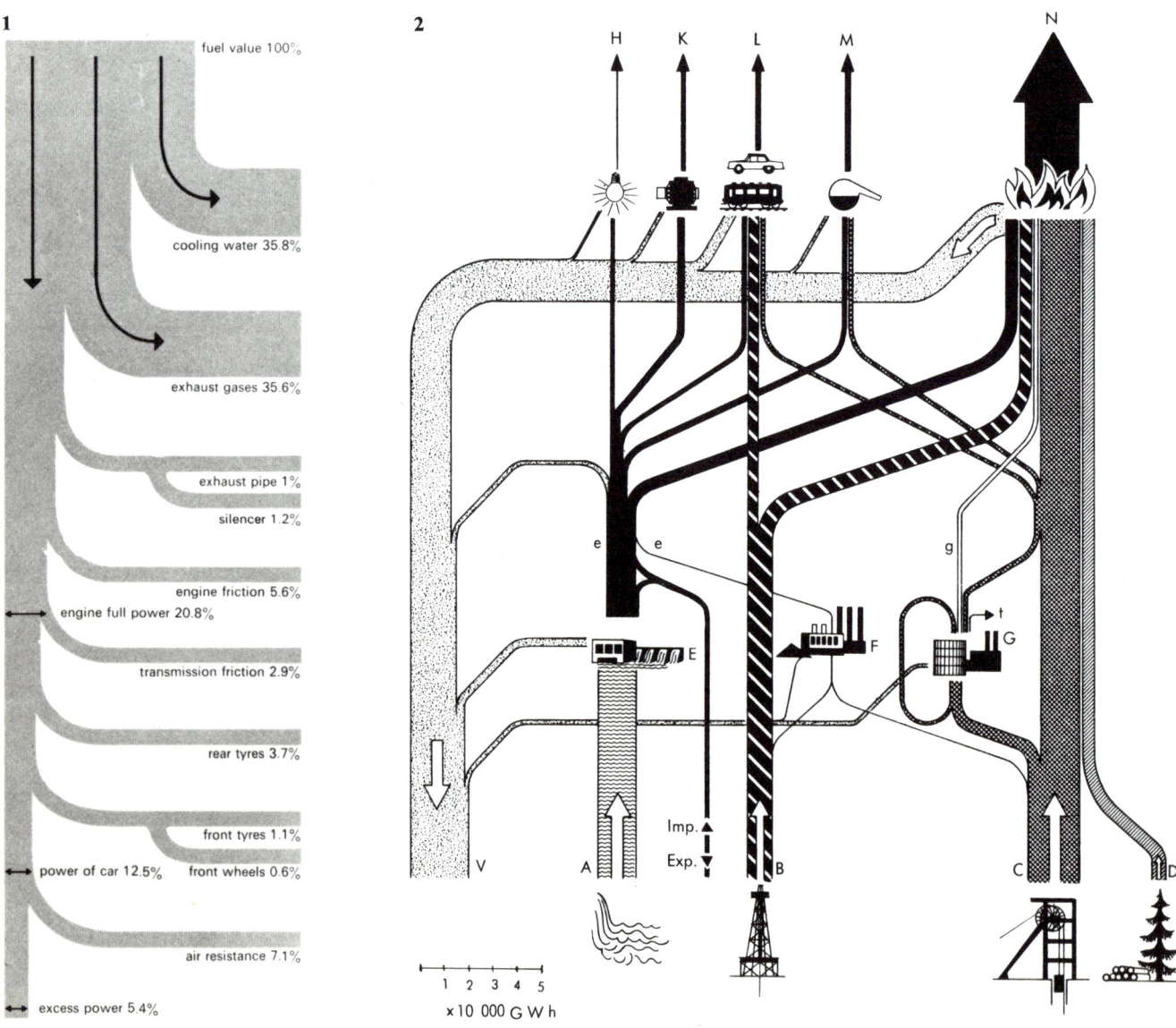

3 Diagram from a text book, showing production of polyvinylchloride (PVC). A. Jentzsch and J. Winkler, *Güterkunde*, Westermann

4 Diagram from a scientific paperback, showing automatic weather station. Line is keyed to describe kind of flow. Michael Overman, *Water*, Aldus Books

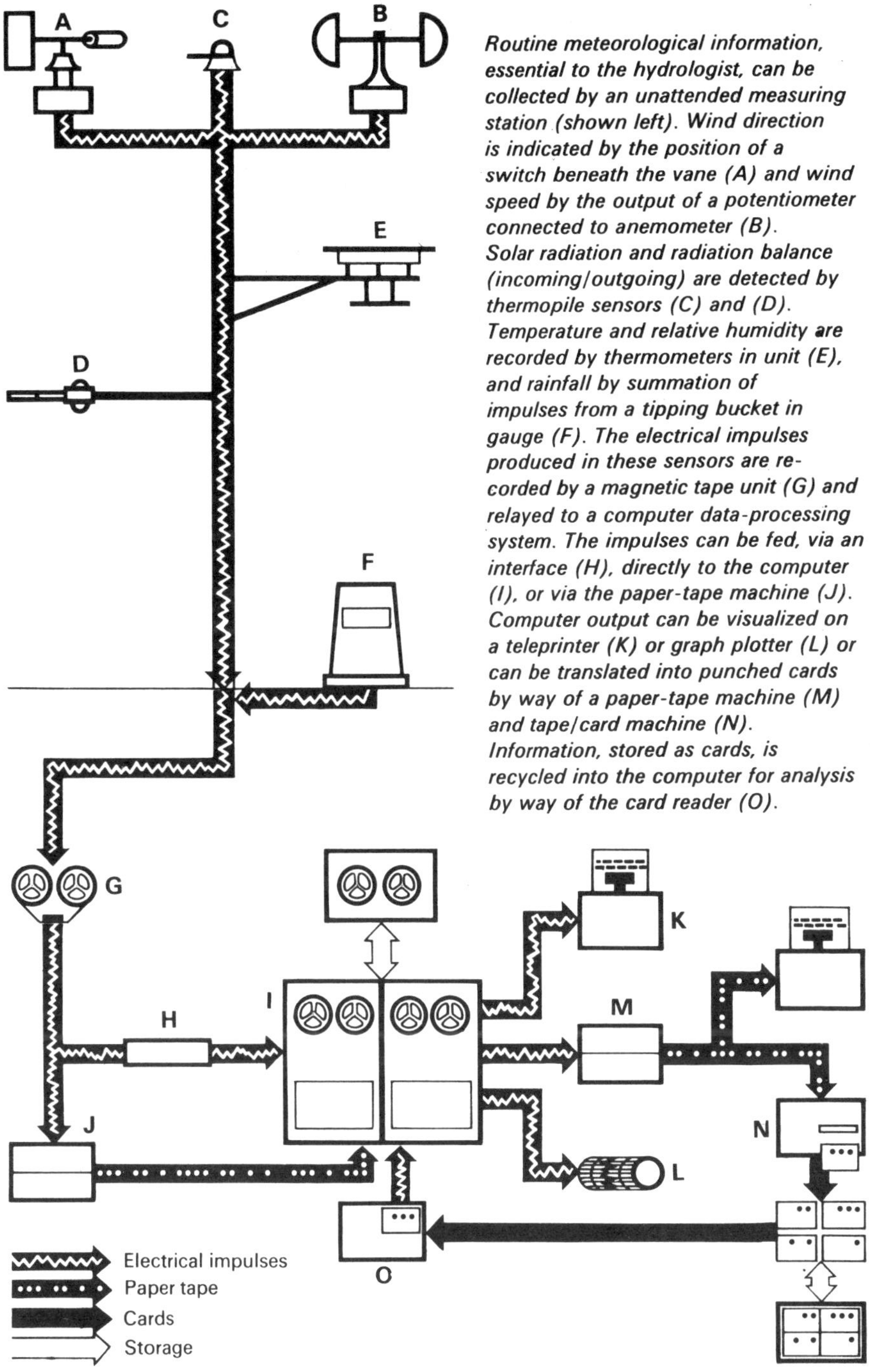

A
C
B
E
D
F
G
H
I
J
K
M
N
L
O
Electrical impulses
Paper tape
Cards
Storage
Routine meteorological information, essential to the hydrologist, can be collected by an unattended measuring station (shown left). Wind direction is indicated by the position of a switch beneath the vane (A) and wind speed by the output of a potentiometer connected to anemometer (B). Solar radiation and radiation balance (incoming/outgoing) are detected by thermopile sensors (C) and (D). Temperature and relative humidity are recorded by thermometers in unit (E), and rainfall by summation of impulses from a tipping bucket in gauge (F). The electrical impulses produced in these sensors are recorded by a magnetic tape unit (G) and relayed to a computer data-processing system. The impulses can be fed, via an interface (H), directly to the computer (I), or via the paper-tape machine (J). Computer output can be visualized on a teleprinter (K) or graph plotter (L) or can be translated into punched cards by way of a paper-tape machine (M) and tape/card machine (N). Information, stored as cards, is recycled into the computer for analysis by way of the card reader (O).

4

3

1
2
3
4
5
6
7
8
9
10
11
12
Polyvinylchlorid

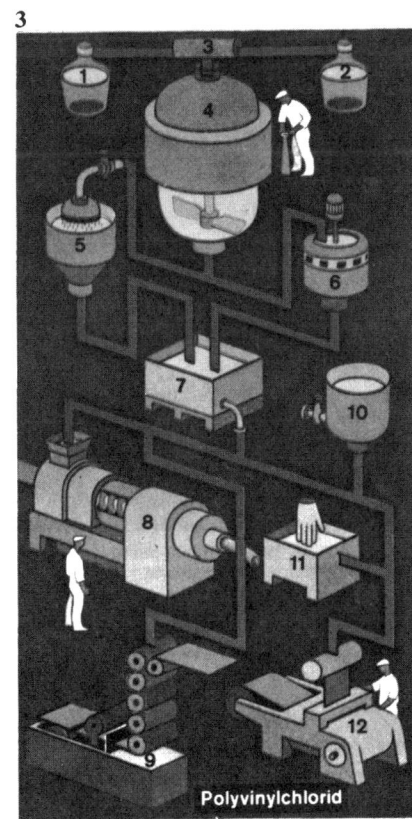

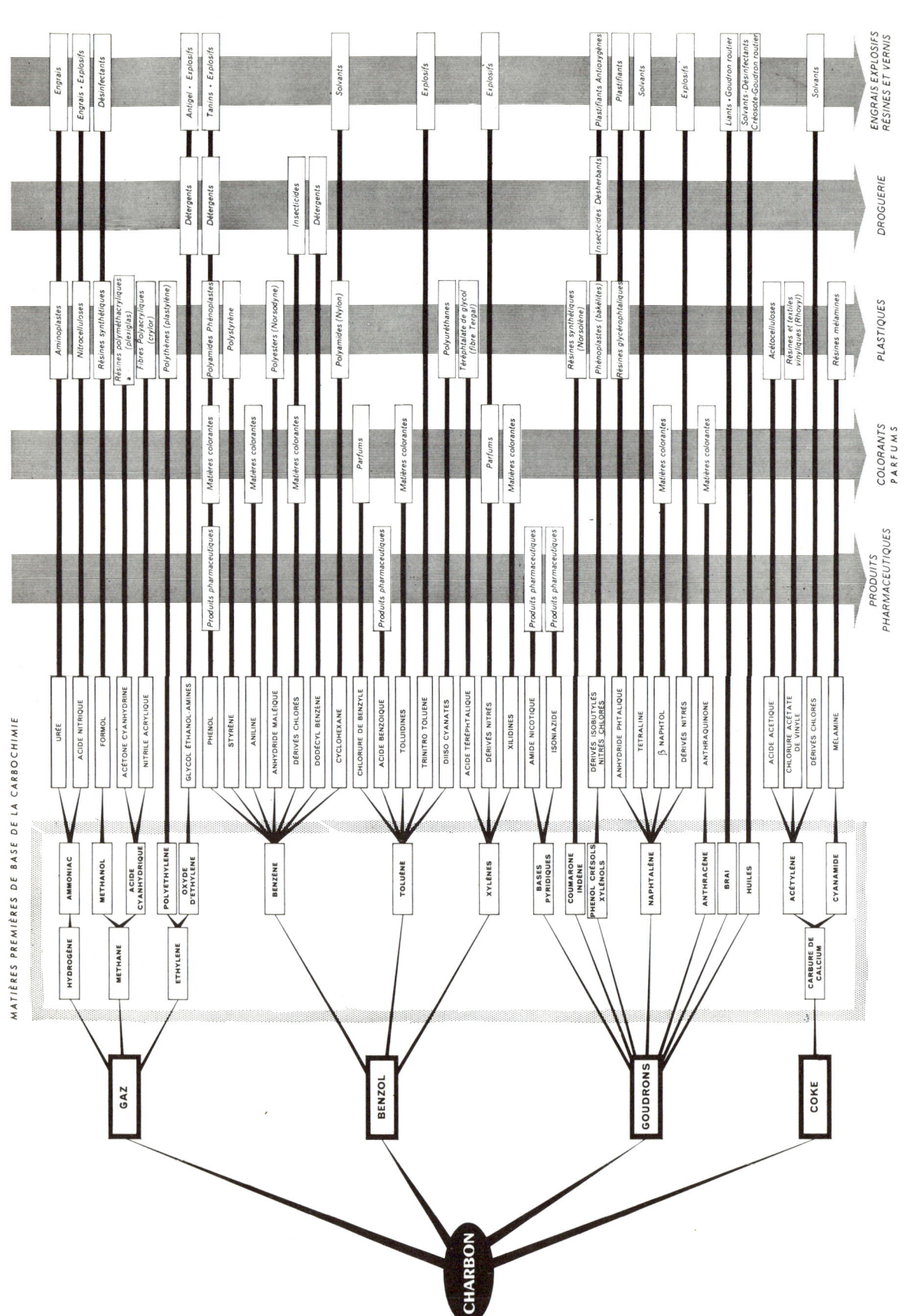

Flow chart

1 Diagram from a school book, showing products derived from coal. *Le Charbon,* Hachette

2 Diagram from a scientific magazine, showing energy flow from sun. See family tree version page 127. *Power*

3 Diagram from a text book, showing stages from ingot to finished product. Compare with typographical method opposite. J. A. Rankin, *Workshop processes and materials for mechanical engineering technicians,* Penguin Education

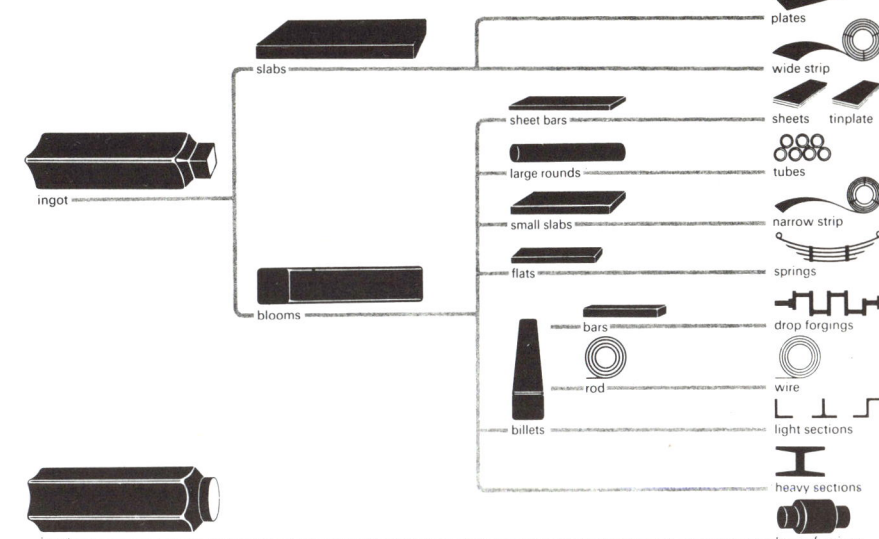

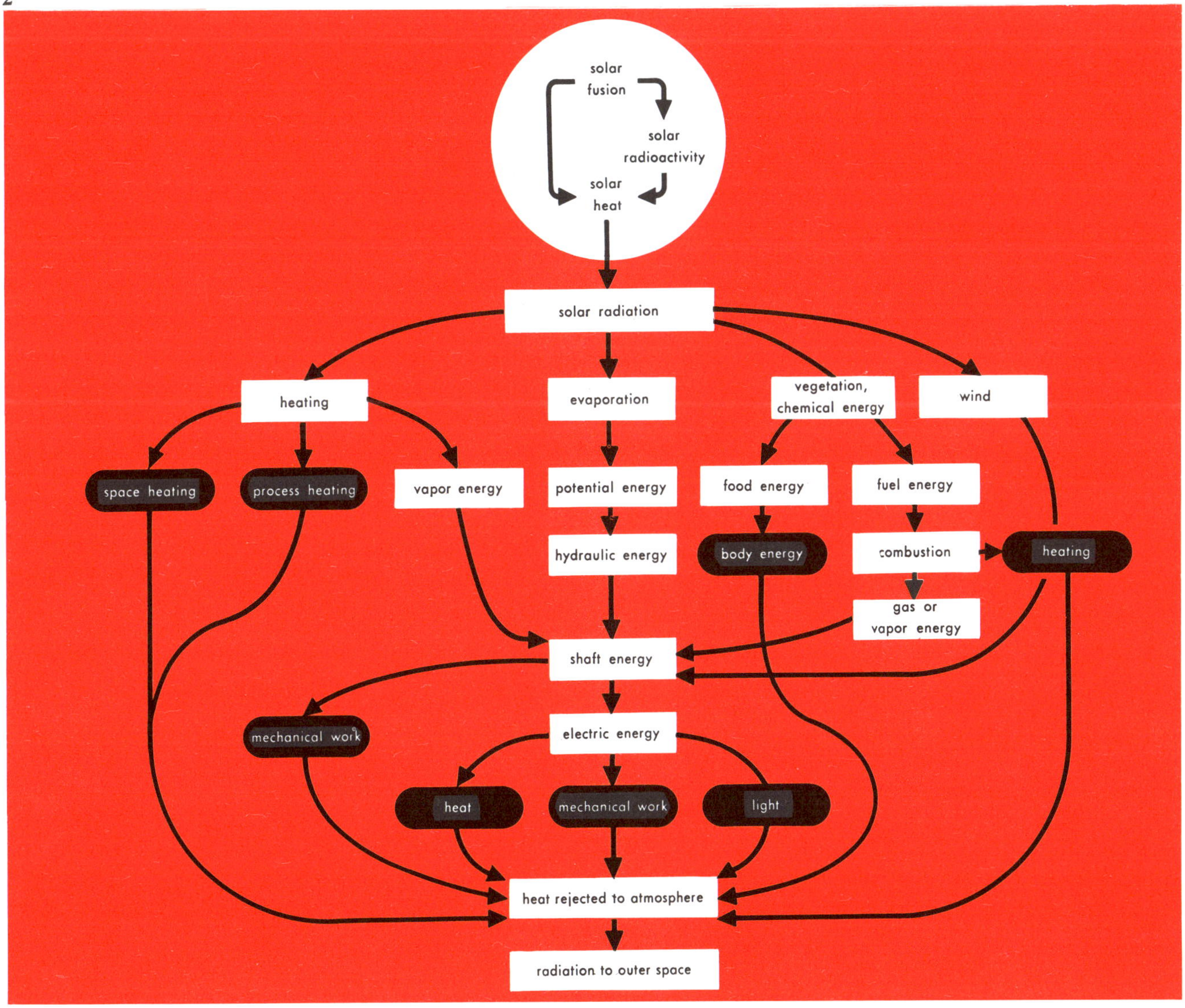

1
Gas from Coal

1   Retort house where coal is baked in ovens
2   Tar being piped off
3   Well for tar storage
4   Primary condensers (main coolers)
5   Exhauster which draws off gas
6   Condenser for further cooling
7   Coke to water-gas plant
8   Coke grading plant
9   Last-of-tar remover
10 Washer where ammonia is removed (by dissolving it in water)
11 Balance holder to control supply of gas
12 Carburetted water-gas from steam coke and oil
13 Purifiers for clean flame and so on
14 Benzene recovery plant
15 To refinery
16 Station meter (or measurer)
17 Gas holder
18 Gas dryer – washing it has left it wet
19 Pressure governor to control supply
20 Street main supply pipes
21 House supply

Flow chart

1   Chart from a text book, showing how gas is made from coal. A modified isometric cut-away section enables information to be shown about particular stages. *Nuffield Chemistry Background Books, Coal,* Longmans/Penguin Books

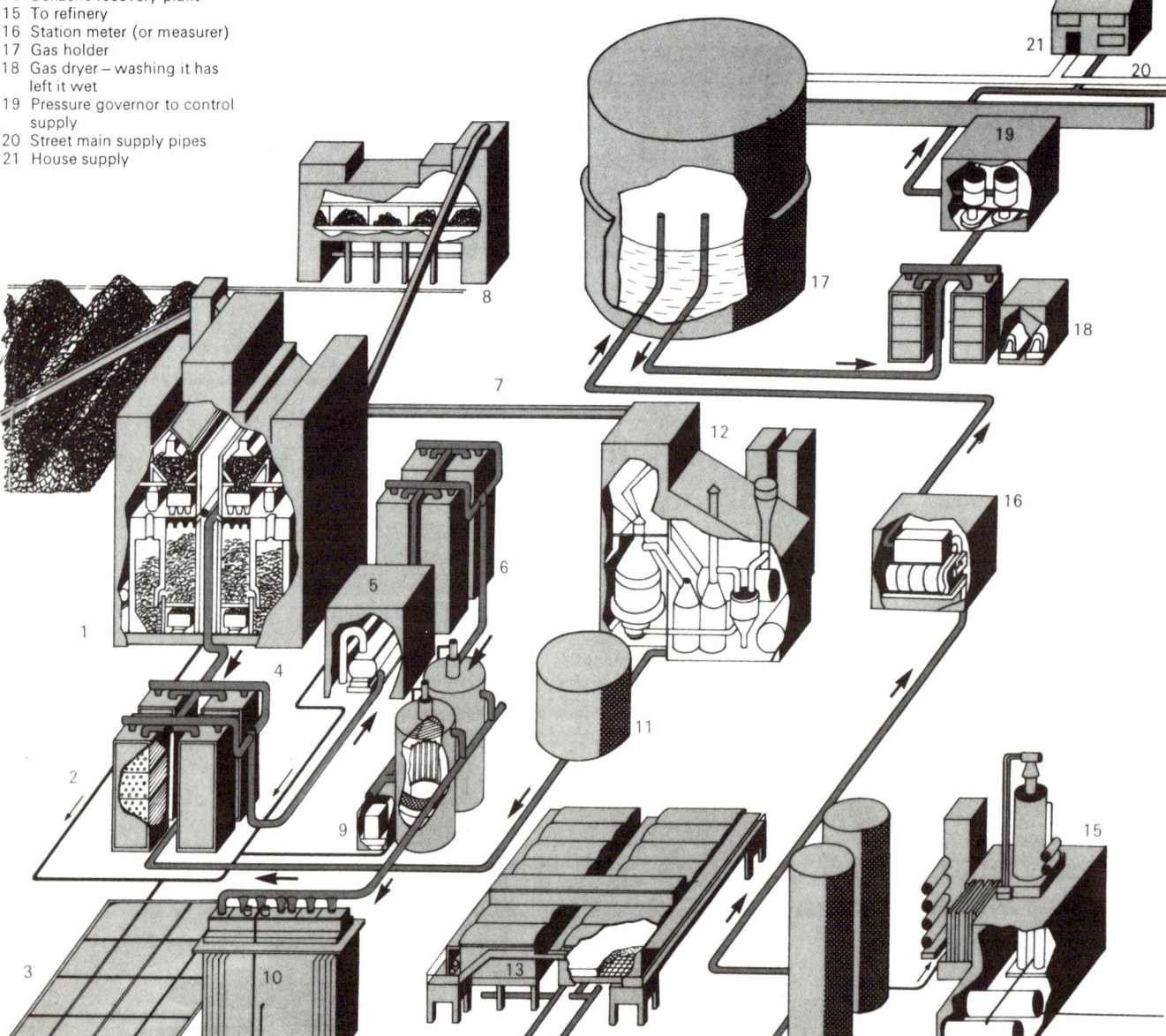

2 Chart from an educational book, showing one computer masterminding the baking of cakes. No attempt is made to represent appearance of machines. Aim of diagram is to show control of various stages by computer. This is done by arrows and long caption superimposed on computer. *Life Science Library, Machines,* Time Inc.

3 Diagram from a paperback history atlas, showing a tading company in the 15th and 16th centuries and its activities. The interrelation of factory, bank, town with trade fair, market town, village etc. Flow lines use symbols to differentiate. *dtv-Atlas zur Weltgeschichte,* Deutscher Taschenbuch Verlag

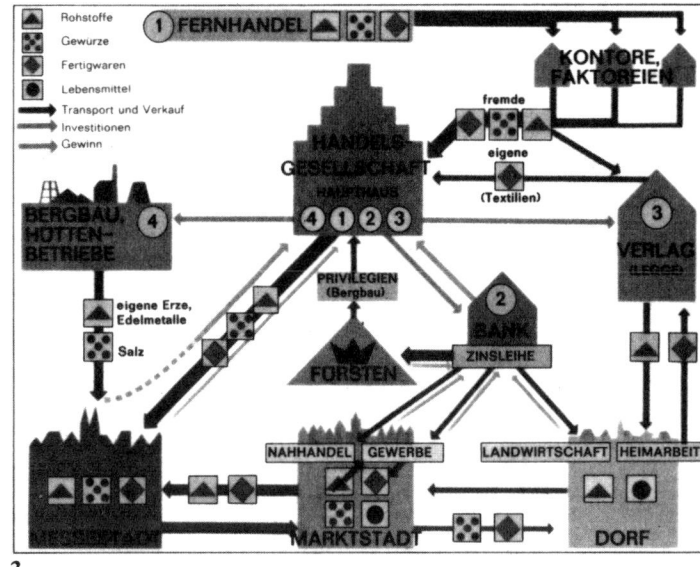

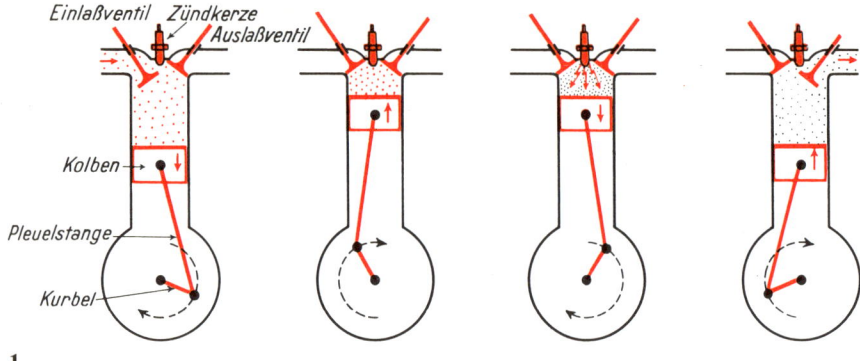

1

The problem for the designer is to produce a diagram which explains a technical process or machine to the layman. According to the audience or aim of the diagram, there is scope for variation in treatment as indicated by just four examples of the diagrams of the internal combustion engine.

1 From a text book. Red used for moving parts and sparking plug. *Physik*, Verlag Moritz Diesterweg

2 From a text book. *Natur und Technik*, Velhagen & Klasing

3 From an educational book. Emphasis on the piston. *Life Science Library, Machines*, Time Inc.

4 From a book for children. One of a series of diagrams explaining the car. Crankshaft left out to simplify explanation. *How it works, The Motor Car*, Wills and Hepworth

5 Diagram from a text book, showing stages in the L.D. process for making steel. Simplification to show basic stages. J. A. Rankin, *Workshop processes and materials for mechanical engineering technicians*, Penguin Education

6 Diagram from a technical publication, showing Pelton turbine. Escher Wyss

7 Diagram from a text book, showing Pelton turbine. Simplification for school audience. Flow of water included to make clear the working of the turbine. *Natur und Technik*, Velhagen & Klasing

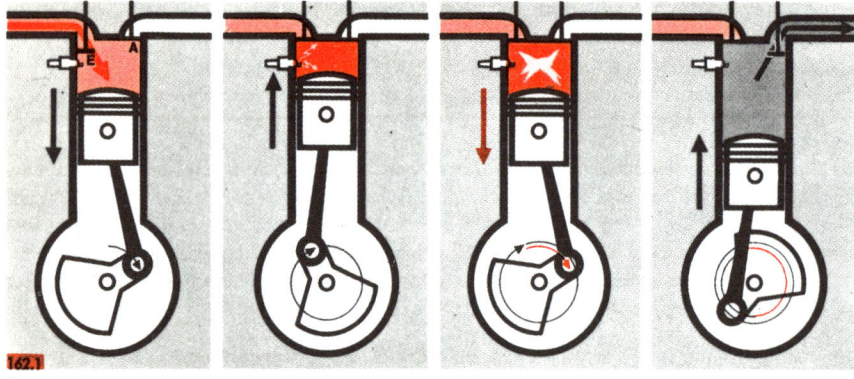

2

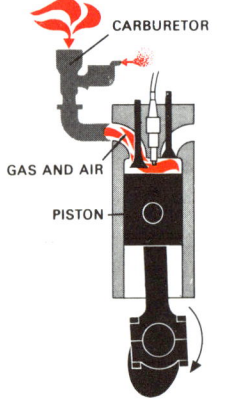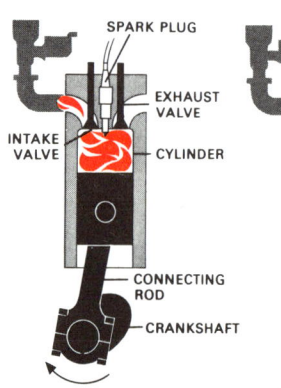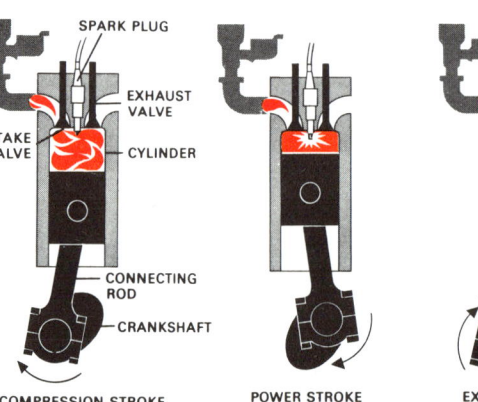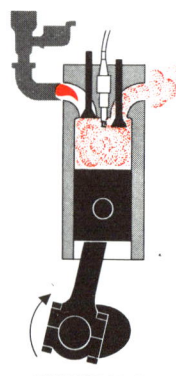

3

4

116

Technical

One of the elements in the flow chart may be a simplified illustration of the machine. This will often be based on a complex drawing done by a technical illustrator.

Technical illustration is a specialist field ranging from a complex cut-away, three-dimensional view of aircraft or engine, to a side elevation of a single screw. The technical illustrator uses certain conventions (briefly described on page 138) which condition the appearance of some of his work. In spite of this, much can be gained by the designer understanding this field and some of the actual techniques of drawing can be of direct benefit (modified isometric projection page 114). The fact that the technical illustration is the source for the diagram done by the designer means that understanding the original is essential in order to simplify.

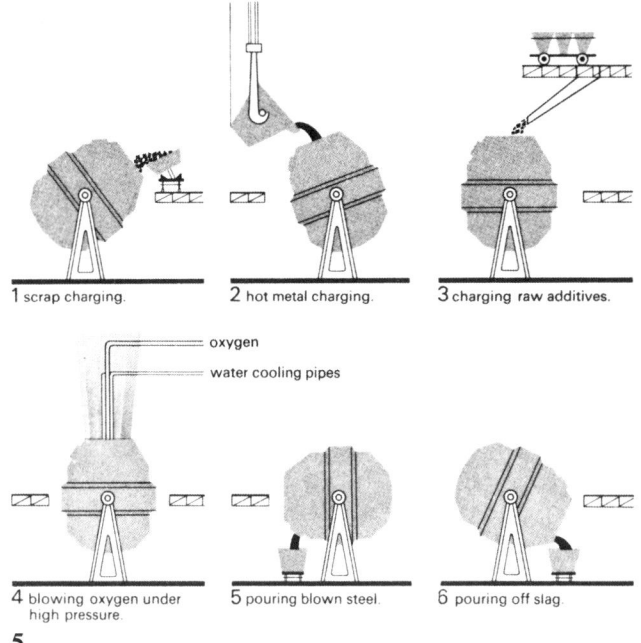

1 scrap charging. 2 hot metal charging. 3 charging raw additives.
4 blowing oxygen under high pressure. 5 pouring blown steel. 6 pouring off slag.

5

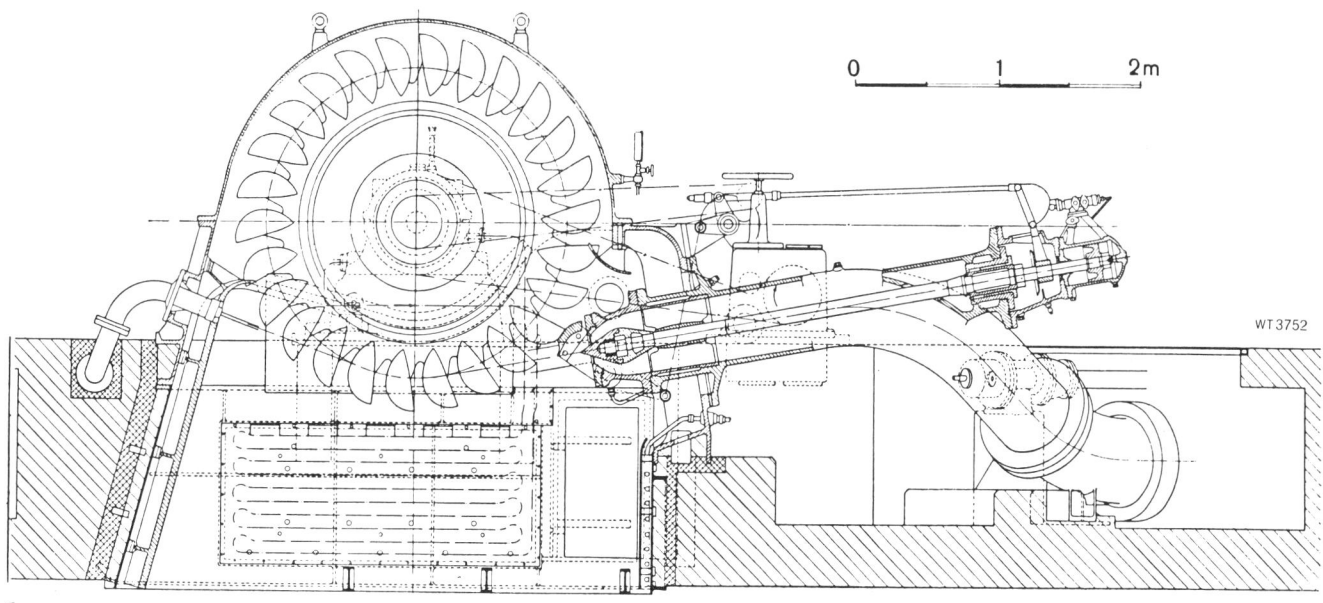

6

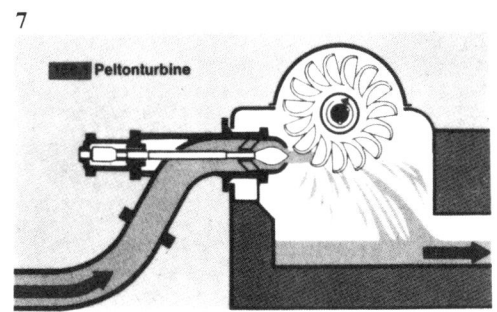

7

Organization chart

The organization chart shows the interrelations, the responsibilities, and the authority of various units of an organization. The units may be officials or departments of a government or business. The names of the units may be enclosed in boxes or circles, which are then joined by a line that shows the delegation of responsibility and authority. The diagram can be purely typographical or it can use symbols. The organization chart does not show quantities. More emphasis can be given to certain boxes by size, weight of line, and choice of typeface.

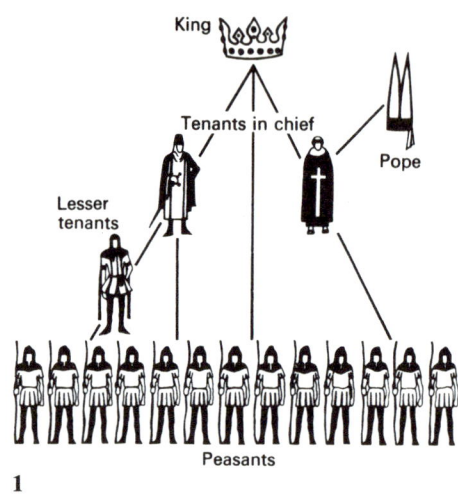

1

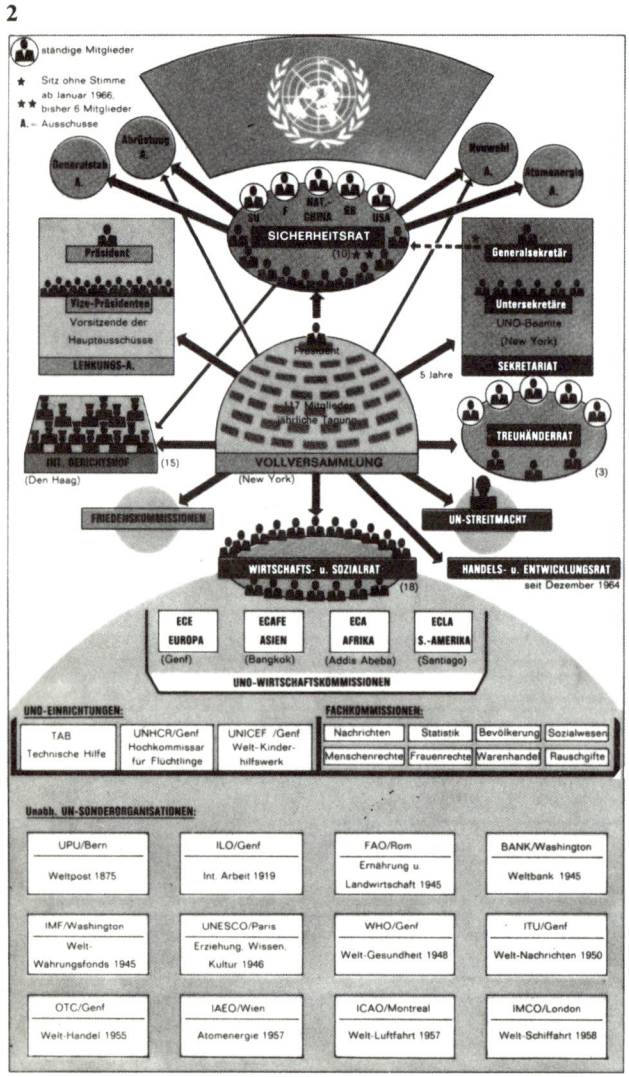

2

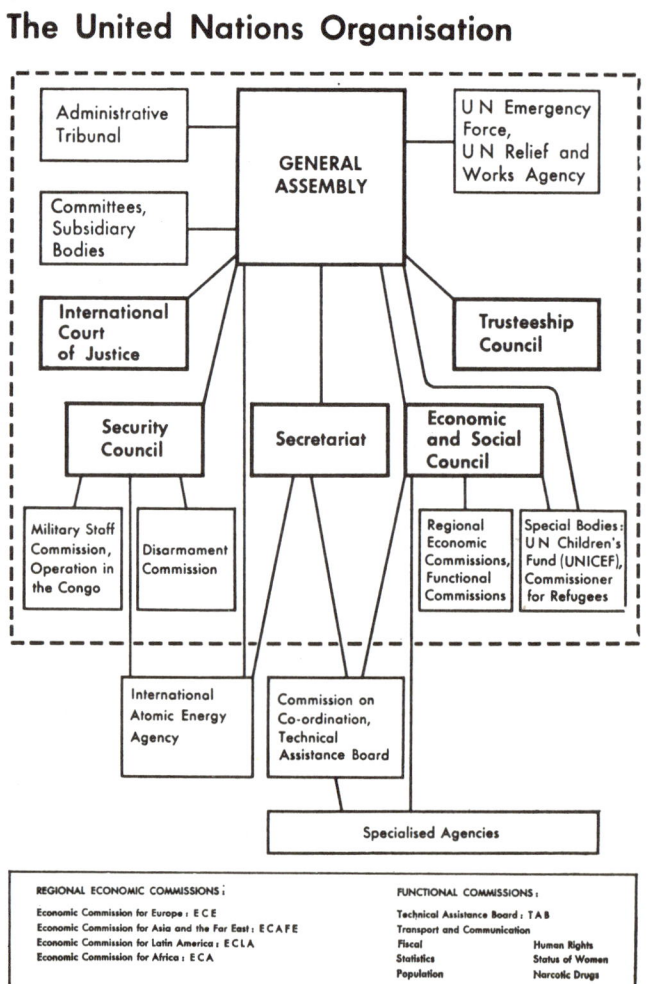

3

The United Nations Organisation

1 Diagram from book, showing structure of medieval society. W. M. S. Russell, *Man, Nature, and History,* Aldus Books

2 Diagram from a paperback history atlas, showing United Nations Organization. Use of symbols, arrows showing flow of authority. *dtv-Atlas zur Weltgeschichte,* Deutscher Taschenbuch Verlag

3 Diagram from a text book, showing United Nations Organization. Typographical. Marie Neurath, *Living with One Another,* Parrish

4 Diagram from a text book, showing United States presidential government. An isotype diagram. Marie Neurath, *Living with One Another,* Parrish

5 Diagram from a paperback history atlas, showing United States government. Original in full colour: legislature (blue), executive (red), judiciary (green). *dtv-Atlas zur Weltgeschichte,* Deutscher Taschenbuch Verlag

6 Diagram from an encyclopedia, showing the interrelation of the American political system which is designed to ensure that the judiciary, legislature and executive can each check and regulate the actions of the other two. The flow diagram has become more dominant than the organization. *Man in Society,* Macdonald

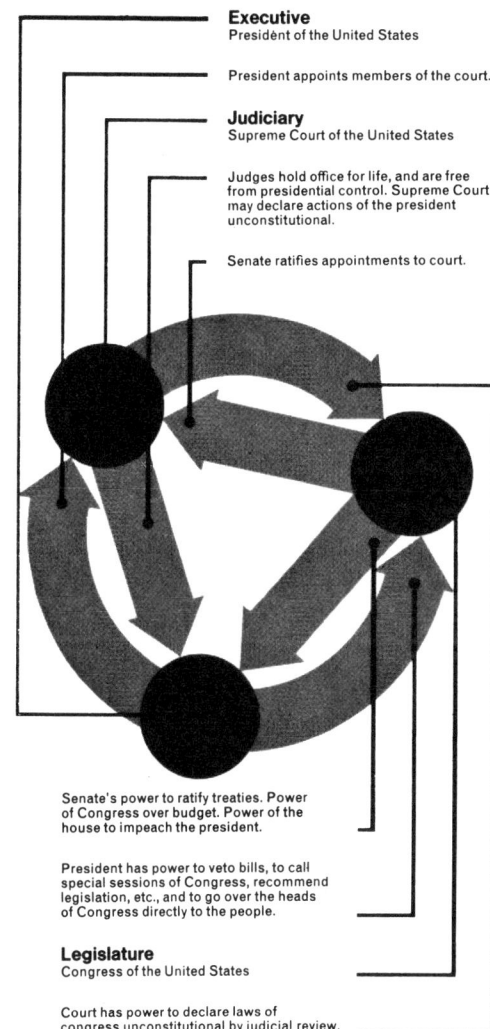

6

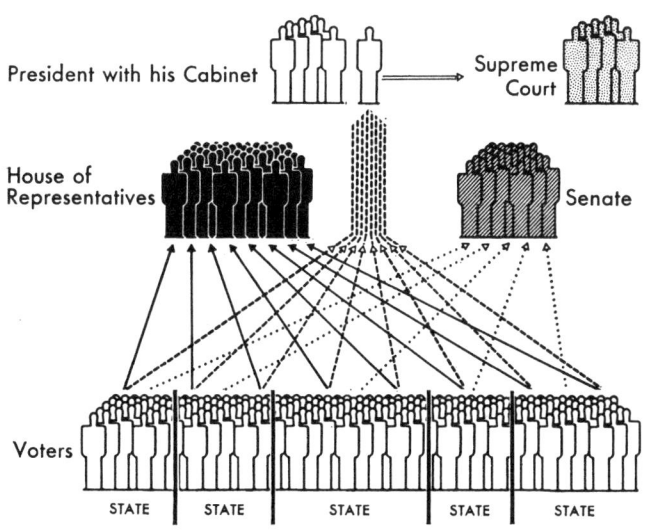

Organization chart

1 Diagram from a business publication, showing the main company and its relations. The Ansul Company

2 Diagram from a paperback encyclopedia, showing classification of flora and fauna. *dtv-Atlas zur Biologie,* Deutscher Taschenbuch Verlag

3 Diagram from business magazine, showing organization of General Motors in 1963. Organization radiating from the President. *Fortune*

4 Diagram from an atlas, showing how education is organized in England and Wales. Visually closely related to the family tree (page 124). *Complete Atlas of the British Isles,* Reader's Digest

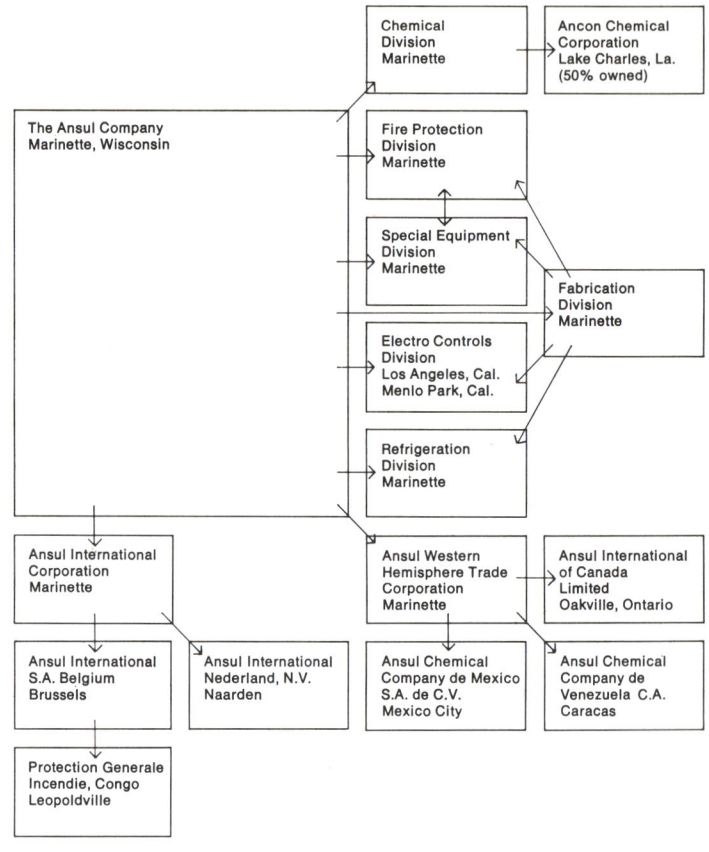

1

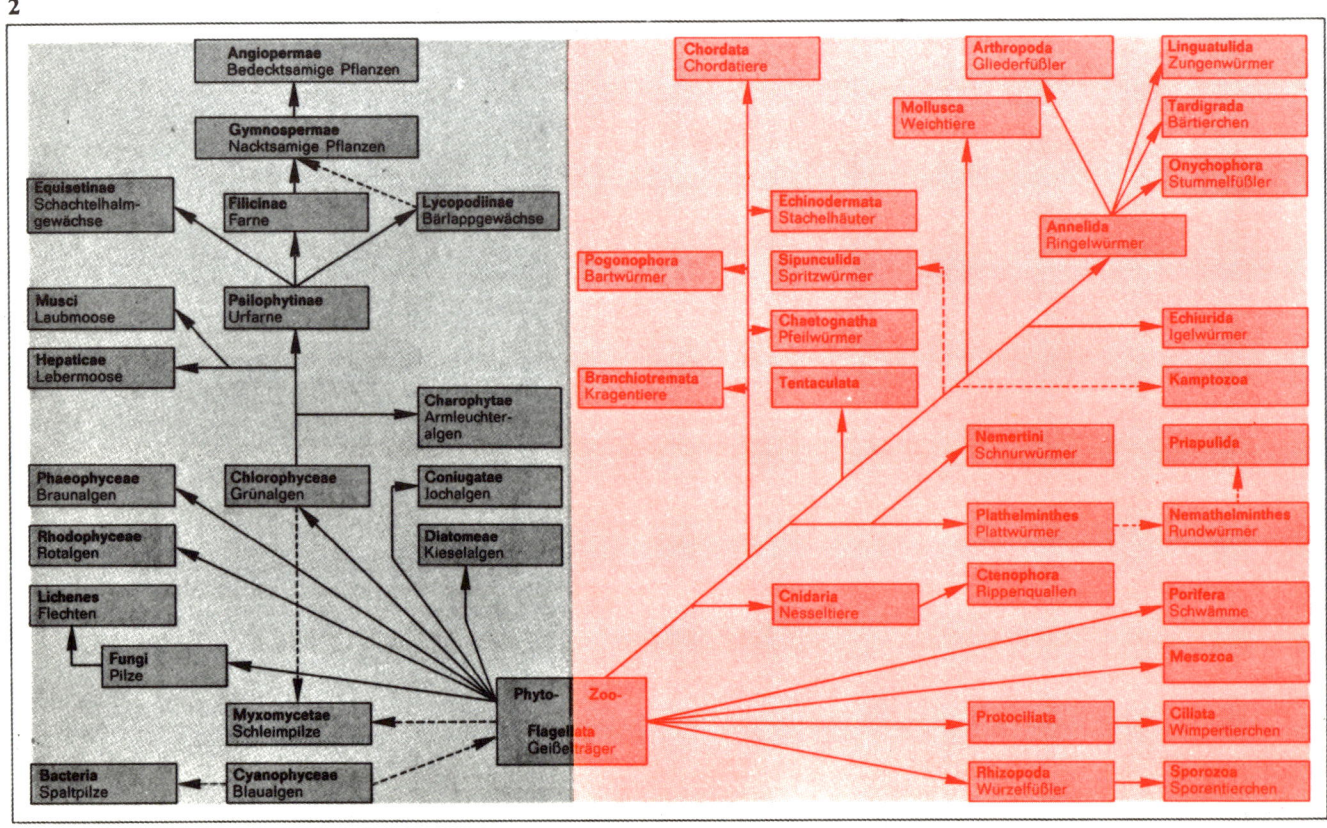

2

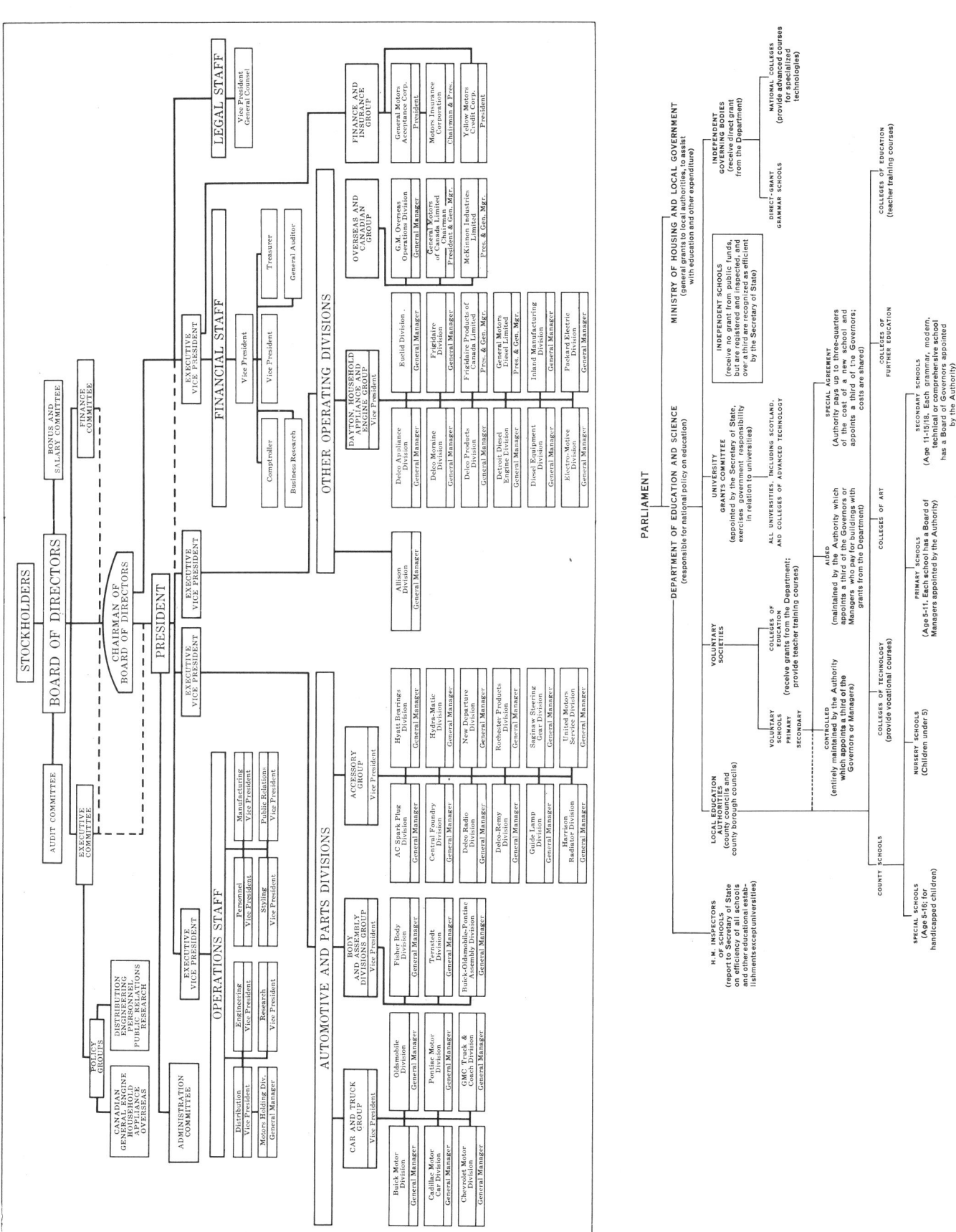

1 Diagram from a book, showing positions for intercourse. A difficult diagram problem ingeniously solved by getting away from direct human representation by use of an artist's wooden model. Man has a light background, woman a black background. *Samlagets Teknik och Variationer,* Orion Press/Wahlström and Widstrand

2 Diagram from a scientific magazine, showing the lungs' gas exchange system. Original full page, in two colours. Elimination of all non-essential information; detail is used only when necessary to tell the story. Cross-section of heart to show flow. *Scientific American*

1

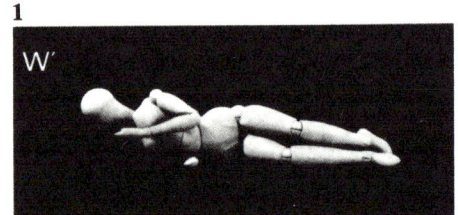

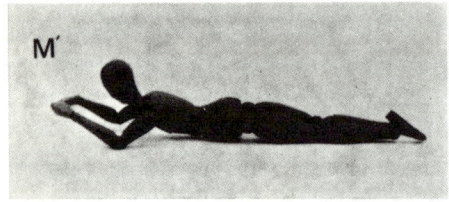

2

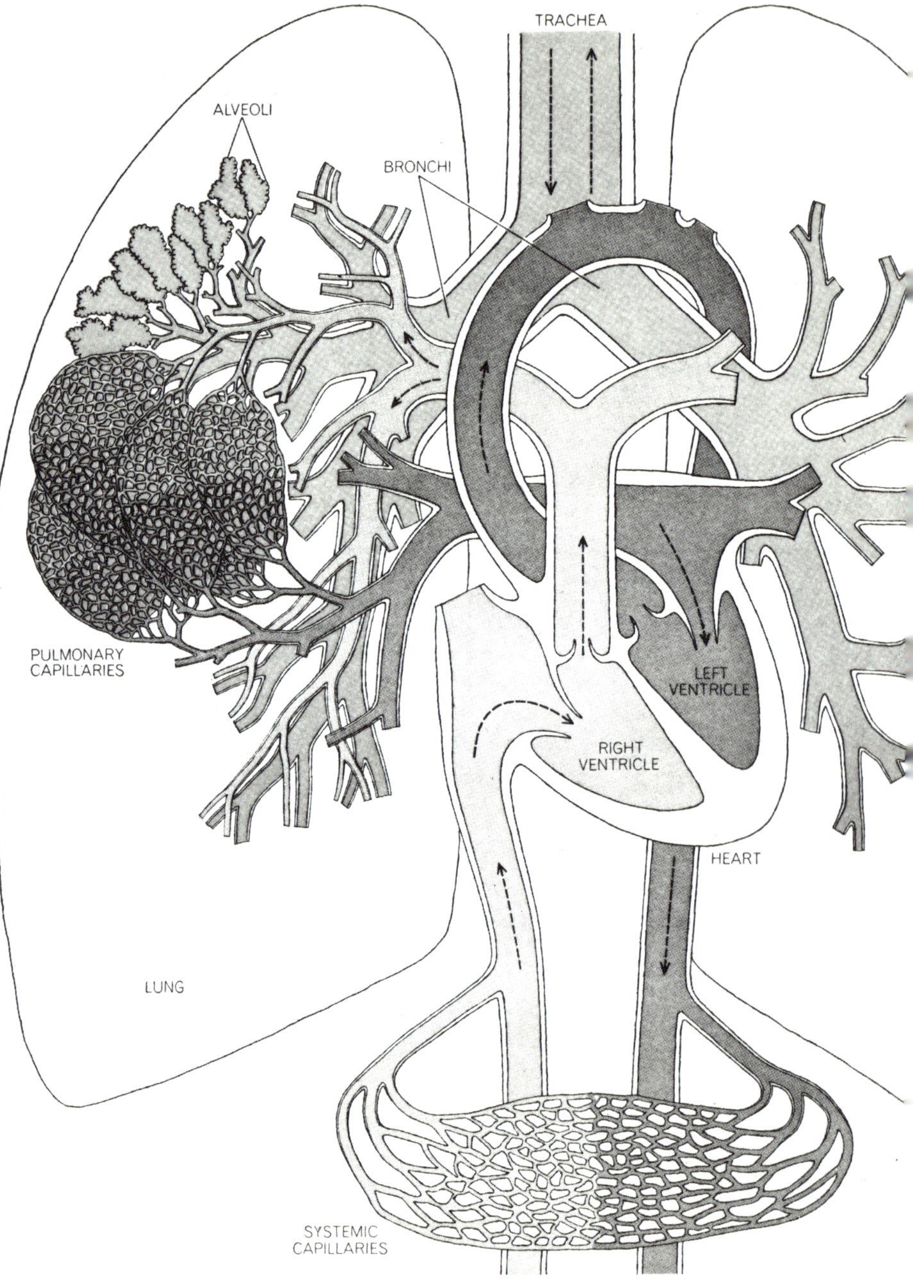

Botanical/medical

The problems here are those of any explanatory diagram simplification in order to get over certain information to a particular audience. The approach required for a scientific magazine, and a school text book, will be quite different in the quantity and complexity of information. The one thing which they will probably have in common is a sensitive technique, carefully thought out line, use of air brush to get soft shading. The hard, bold lines of the technical diagram seem out of place in this field.

3 Two pages from a paperback biology encyclopedia. Pages of diagrams in full colour face pages of relevant text. The approach and quality of the diagrams is exceptional. The line work is sensitive, the labelling clearly organized, the colour carefully controlled on each page. *dtv-Atlas zur Biologie,* Deutscher Taschenbuch Verlag

3

82 Organe / Organsysteme der Wirbeltiere I: Haut, Atmungssystem

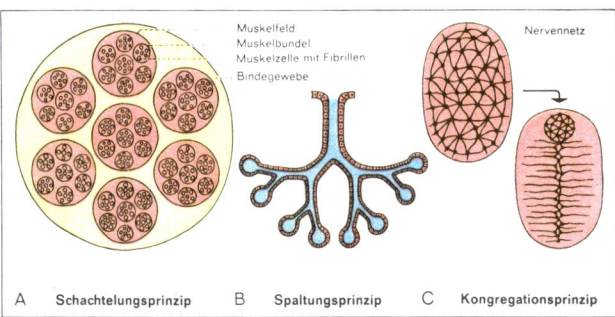

Prinzipien der Organbildung

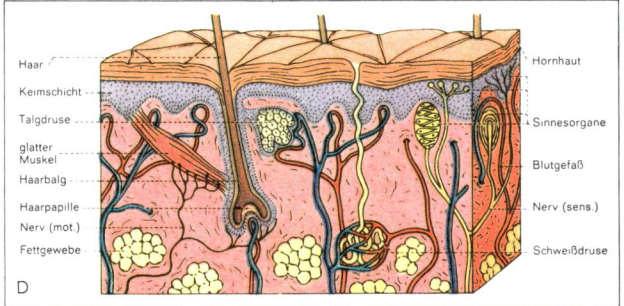

Bau der menschlichen Haut

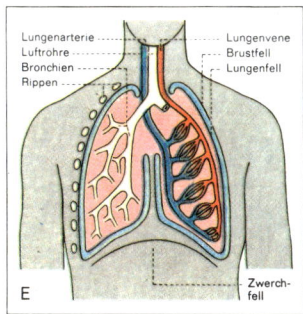

Lunge: Bronchienverzweigung und Gefäßversorgung

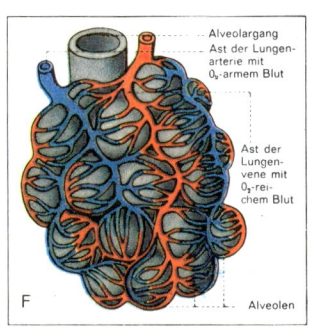

Alveolensäckchen mit Gefäßversorgung

92 Grundtypen der Lebewesen / Baupläne der Kormophyten I: Blattstellung, Sproßverzweigung

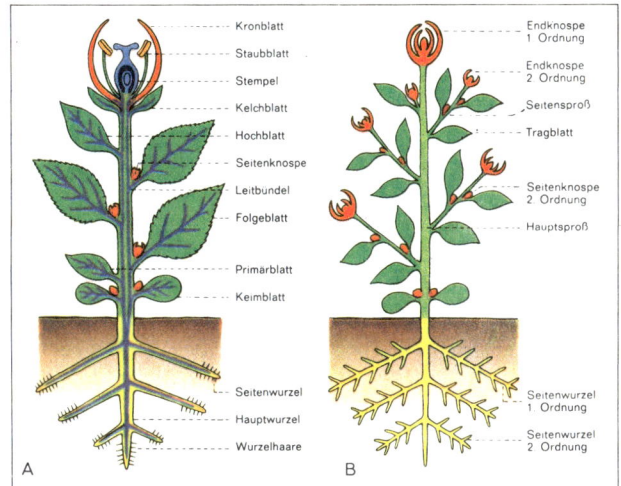

Baupläne von Kormophyten: mit Blüte (A), mit verzweigtem Sproß- und Wurzelsystem (B)

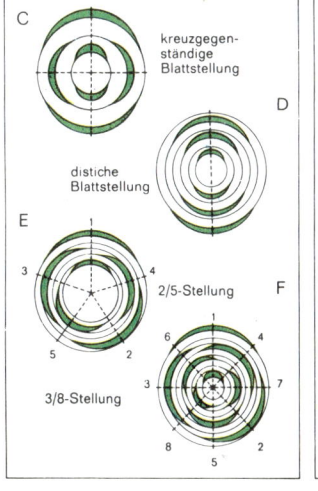

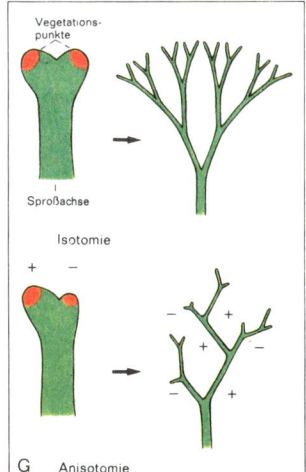

Blattstellungstypen

Dichotome Verzweigung

123

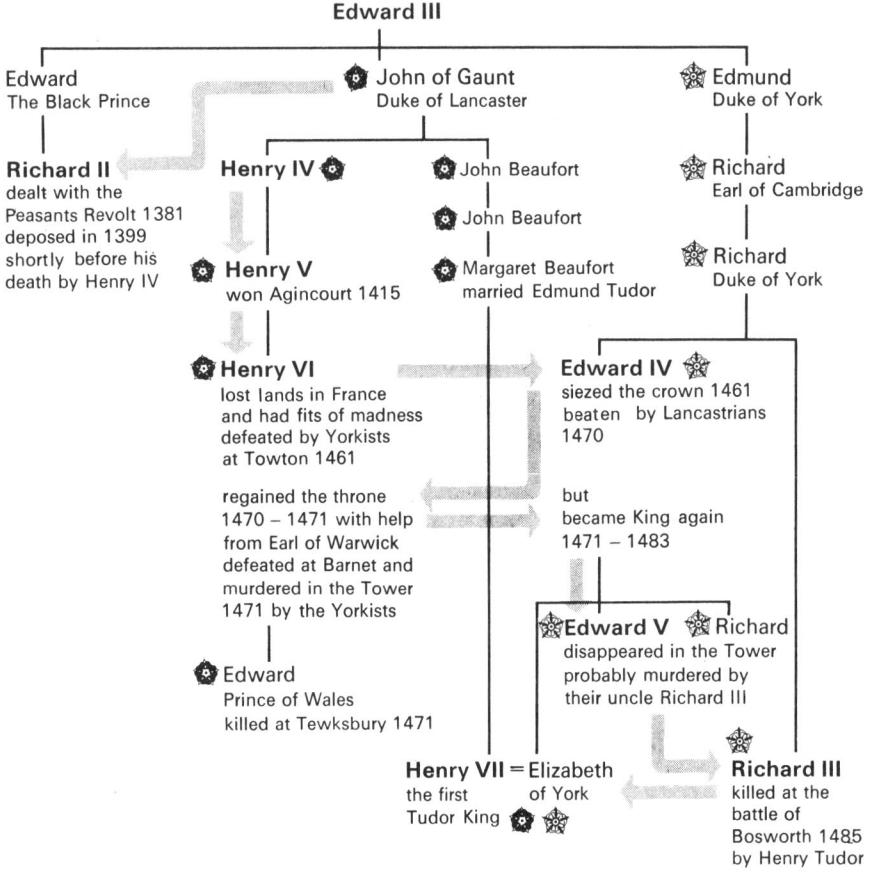

1   Diagram from a text book, showing the families of Lancaster and York. Light (York) or dark (Lancaster) roses indicate which family, bold type indicates king. Valerie E. Chancellor, *Medieval and Tudor Britain,* Penguin Education

2   Diagram from an encyclopedia, using symbols. Original colour coded to show inheritance factor. *Nature,* Macdonald

3   Diagram from a text book, showing part of the family tree of the Wedgwood, Darwin and Galton families. Solid squares mark men who became Fellows of the Royal Society. *Nuffield Biology Text V,* Longmans/Penguin Books

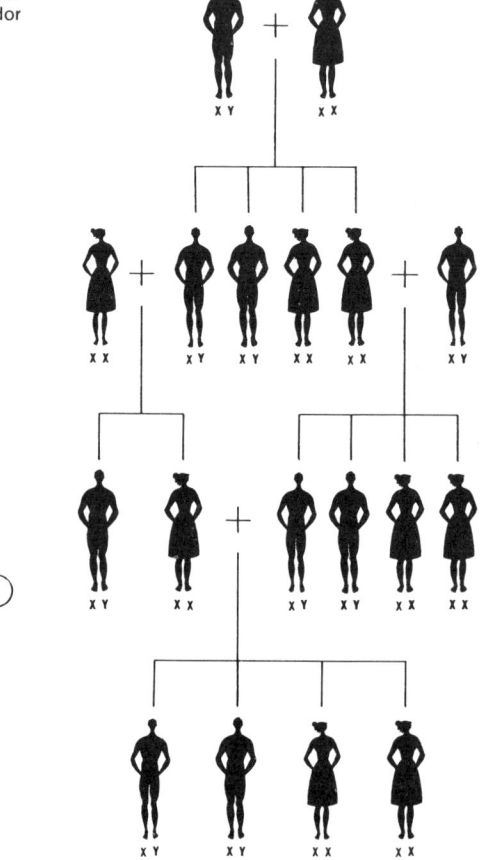

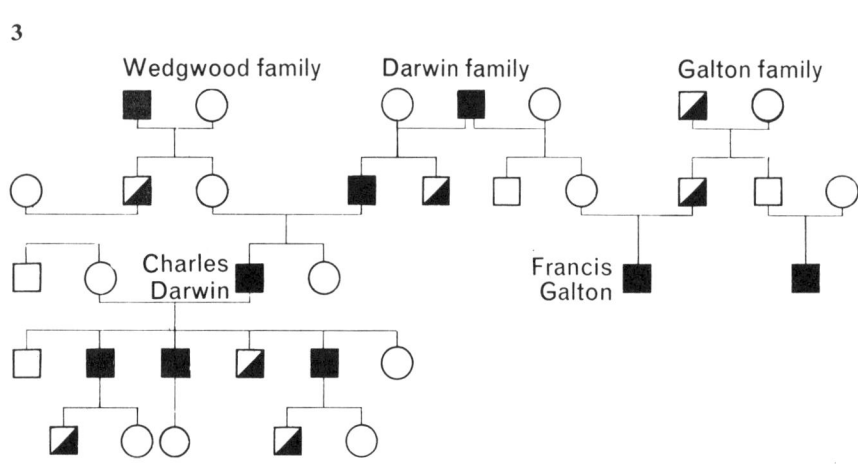

Family tree

Family relationships are shown on a diagram commonly known as the family tree (or genealogical, genetic, ancestral chart). It is usually designed to trace the genealogy of one particular person, and can be typographical or may use representational or abstract symbols.

4 Diagram from a newspaper, showing the Royal family tree. This ingenious solution giving the ancestors of two people, does not show brothers or sisters, only parents. *Sunday Times*

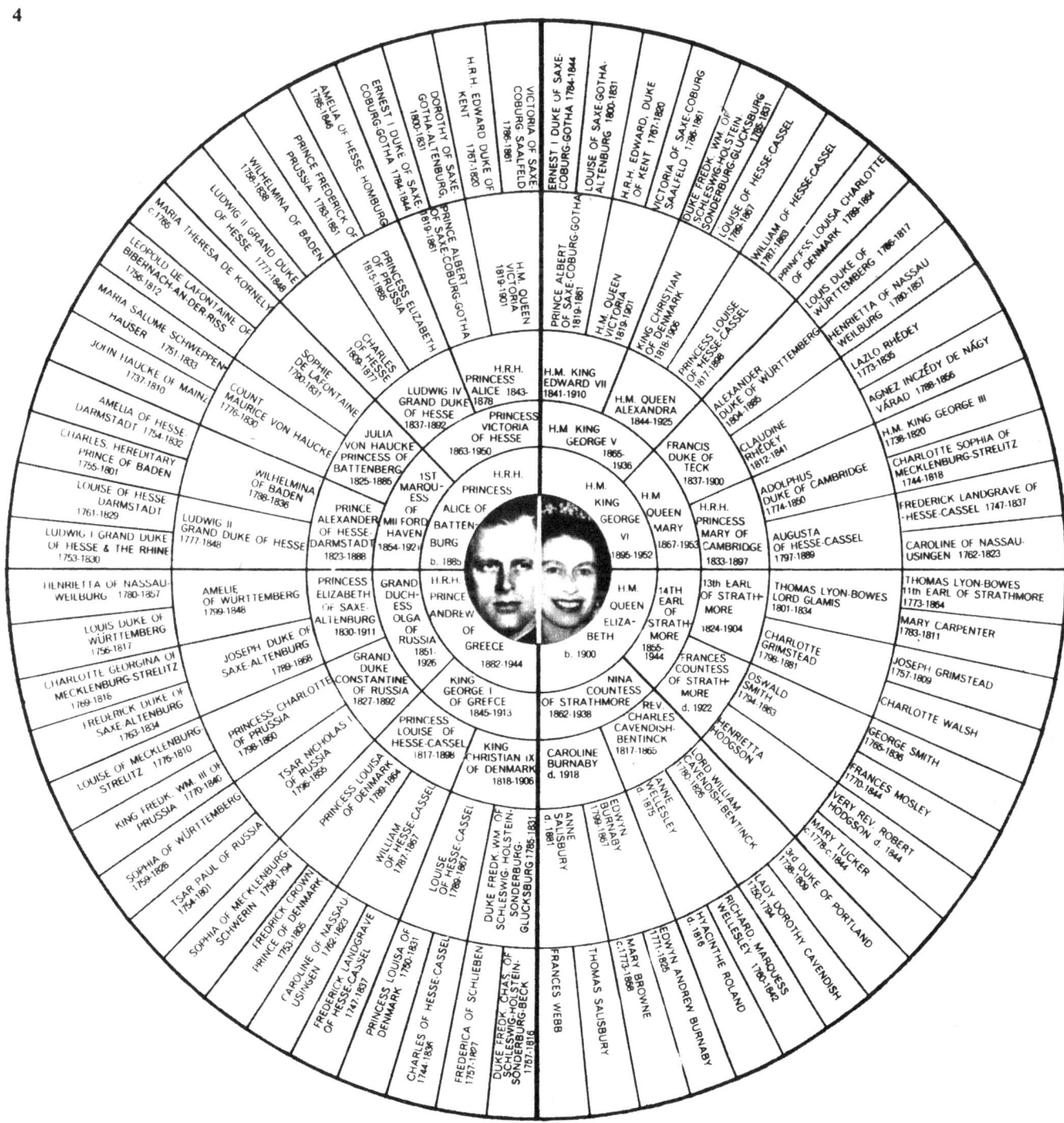

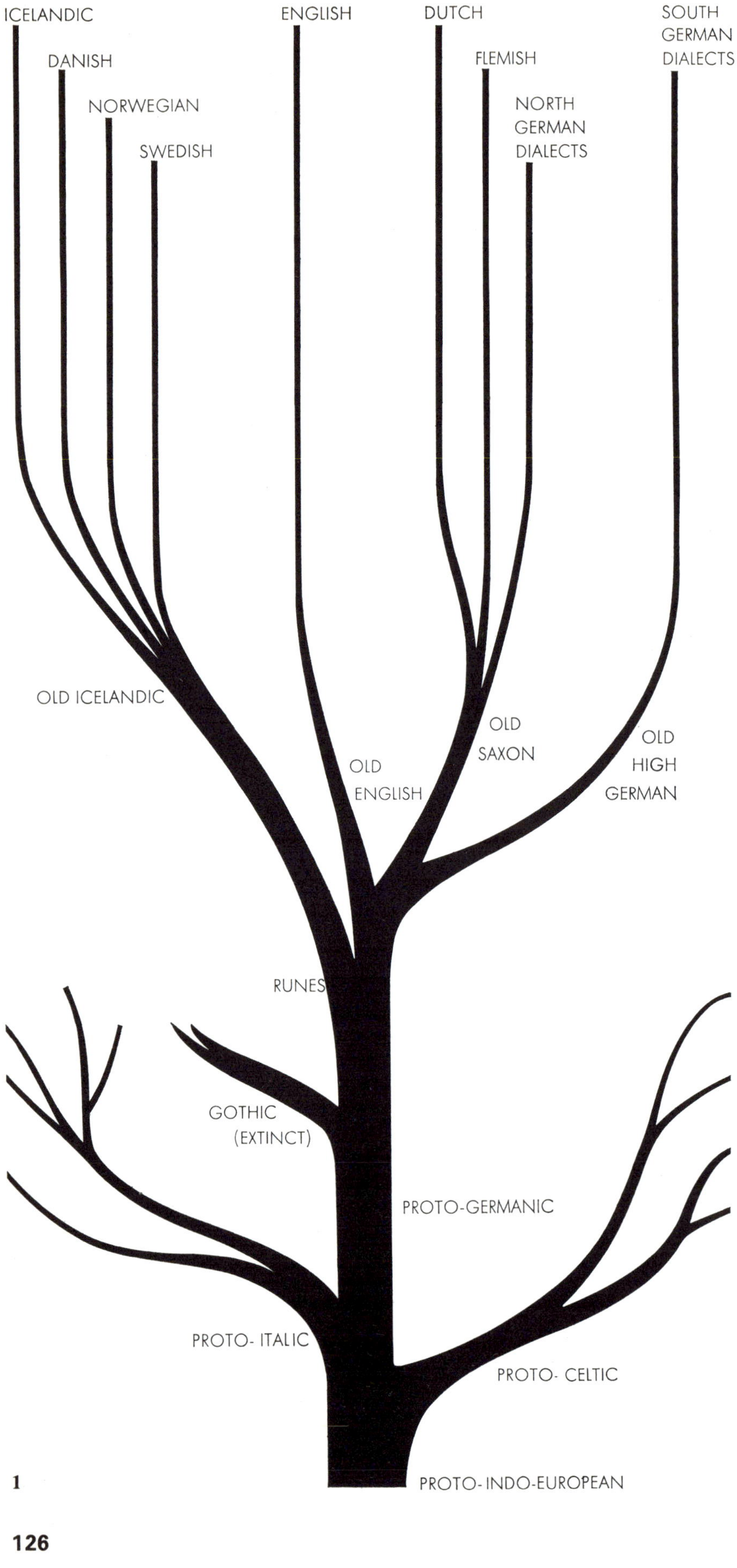

1

ICELANDIC DANISH NORWEGIAN SWEDISH ENGLISH DUTCH FLEMISH NORTH GERMAN DIALECTS SOUTH GERMAN DIALECTS

OLD ICELANDIC
OLD ENGLISH
OLD SAXON
OLD HIGH GERMAN
RUNES
GOTHIC (EXTINCT)
PROTO-GERMANIC
PROTO-ITALIC
PROTO-CELTIC
PROTO-INDO-EUROPEAN

2

'The earliest age at which a woman can draw a retirement pension is 60. On her own insurance she can get a pension when she reaches that age, if she has then retired from regular employment. Otherwise she has to wait until she retires or reaches age 65. At age 65 pension can be paid irrespective of retirement. On her husband's insurance, however, she cannot get a pension, even though she is over 60, until he has reached age 65 and retired from regular employment, or until he is 70 if he does not retire before reaching that age.'

Family tree

The family tree method has been adapted to show relationships outside the human family – languages, energy sources, evolution etc.

1 Diagram from scientific magazine, showing origin of modern Germanic languages. *Scientific American*

2 Diagram devised by Dr Peter Wason and Dr Sheila Jones, demonstrating how continuous prose in Government pamphlets can be abandoned and replaced by a diagram. It is based on the 'logical tree' method. This method leads the reader through successive stages, enabling him to read only what is relevant. The reader has only to decide 'yes' or 'no' at each stage. *Observer*

3 Diagram from a newspaper, showing man's family tree. *Observer*

4 Diagram from an educational book, showing 'family tree' of energy sources and of ways of converting the energy into power. E. G. Sterland, *Energy into Power*, Aldus Books

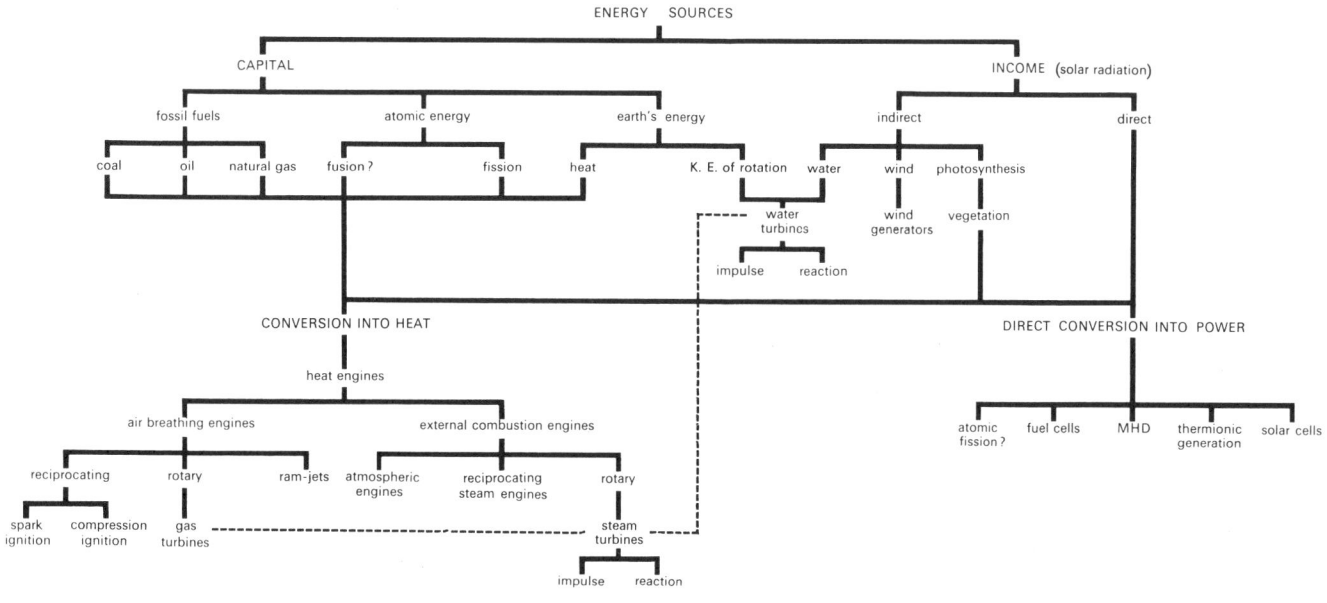

Tabulation

An arrangement of figures or facts in list form. At its simplest, purely typographical; more complicated tables use symbols. Qualitative statements can be made by the number of symbols used.

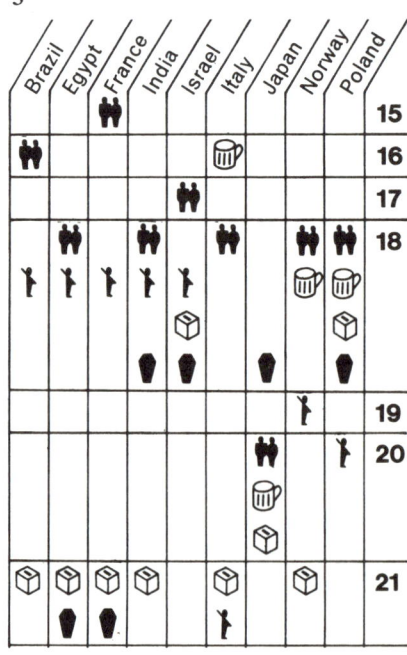

1 Diagram from a motoring atlas, showing the mileage between various towns in southern Britain. *The Reader's Digest AA Book of the Road*

2 Diagram from a scientific magazine, showing research programmes carried out in Antarctica in 1962 listed by nation and field of study. Effective use of flags. *Scientific American*

3 Diagram from an encyclopedia, showing the ages in various countries at which men are liable for military service (soldier), permitted by law to marry (man and woman), may vote (ballot box), may drink in licensed premises (glass), may be executed for capital offences (coffin). *Man in Society*, Macdonald

4 Part of a diagram from a scientific magazine, showing major research in the energy field that could lead to technological improvement (marked with a dot). *Science Journal*

5 Diagram from a consumer magazine, showing the rating of sunglasses. More symbols mean better glasses. *Which?*

6 Part of a diagram from a scientific house-magazine, showing some fields of application for dyes suitable for polyamide. *Ciba Review*

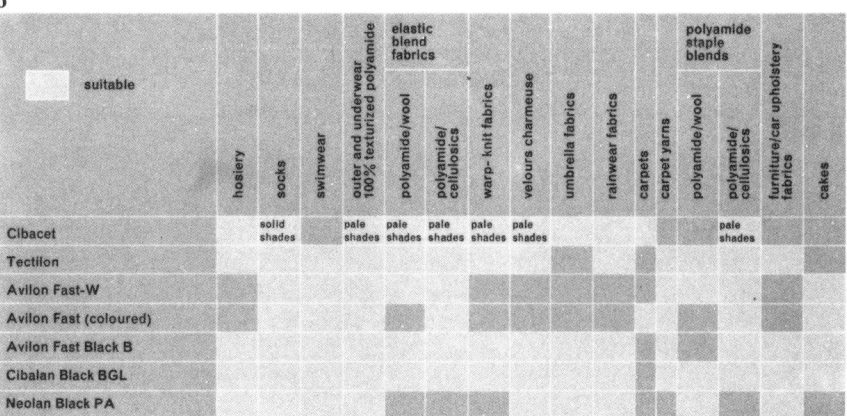

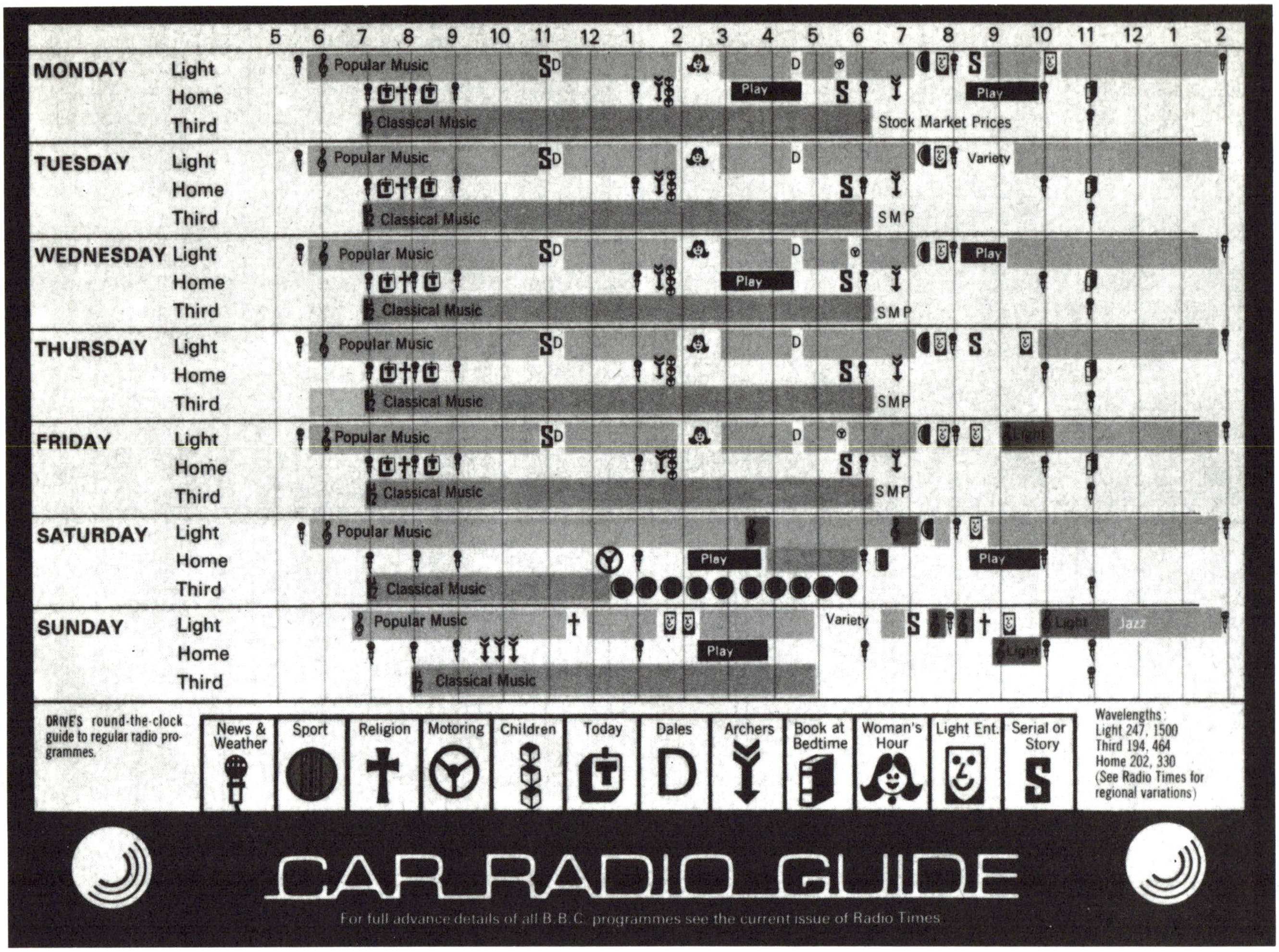

1

2

Time chart

This diagram shows information related to a time scale. The time period covered may be a day (divided into hours), a year (divided into months), recorded history (divided into hundreds of years). With historical time charts the main problem is the small amount of information to be shown for the early part and the overcrowding for recent history. It is difficult to overcome this, but if the chart is shown in separate parts of a book, the simplest method is to change the time scale. This must be clearly labelled to avoid a misleading interpretation of information.

Quantitative statements can be made on such time charts. They can also show spheres of influence or areas covered by, for example, empires or civilizations.

1 Chart from a motoring magazine, showing radio programmes for a week. Original in full colour. Bands of different colour show popular, classical, light music, jazz and plays. Symbols are used for news, sport, religion etc. A concise summary – approximately the same area is needed to show typographically in a newspaper the programmes for the day. *Drive*, Automobile Association

2 Diagram from an educational book, showing the life history of a Japanese beetle. L. Hugh Newman, *Man and Insects*, Aldus Books

3 Chart from an encyclopedia, showing lives of artists living between 1700 and 1750. One of a series of charts. Bars indicate the life span. It is easy to see who were contemporaries in 1700 and 1750, but lack of a vertical scale makes it less easy for intermediate dates. *Man the Artist*, Macdonald

Time chart

1 Chart from a book, showing chief schools of European painting from Giotto. Devised by Eric Newton who explains that it attempts to indicate relative importance of schools (area of shaded masses), approximate dates, principal artists (circle), relative importance (size of circle), threads of influence between schools and artists. Eric Newton, *European Painting and Sculpture,* Faber and Faber

2 Chart from a text book, showing empires. Original in full colour. Each empire is shown by area as well as by time. Thus the Roman empire occupied the mediterranean area, part of Europe, and part of the near East and central Asia. The British empire occupied part of America, the near East and central Asia and part of the far East. An Isotype diagram. *Living in the World,* Parrish

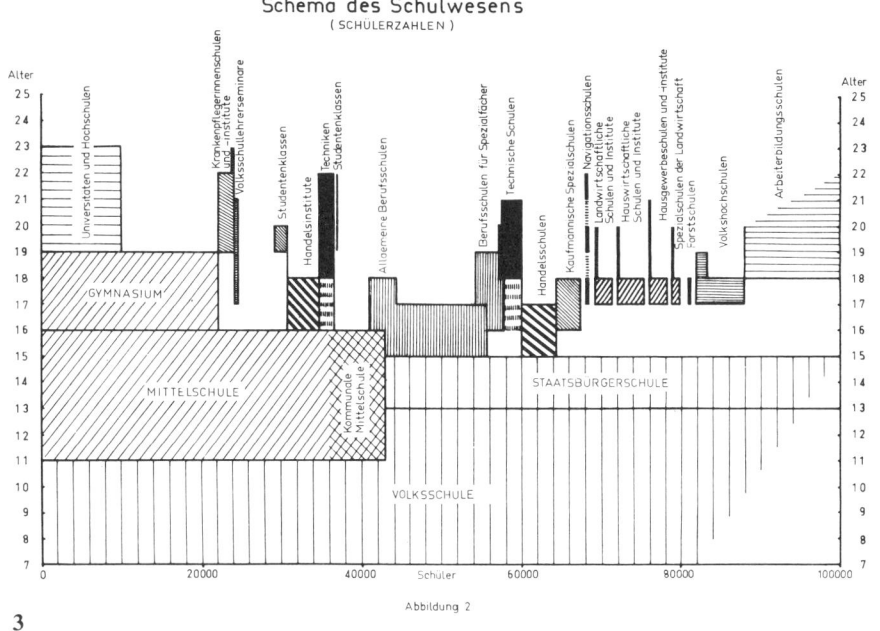

3 Chart from a government publication, showing the school system in Finland – the number of children going to each particular type of school and the age at which they go. A quantitative time chart. Aarno Niini, *Berufsausbildung*

4 Chart from a text book, showing when native fruit and vegetables of a region were taken to another part of the world. There are four regions, and twelve plants (maize, potato, tomato, apple, wheat, grapes, olives, rice, sugar cane, orange, peach, and sugar beet). The time scale is simple, early, Roman, Arabian, after 1500, today. Time, area, and flow combined on one chart. An Isotype diagram. *Living in the World*, Parrish

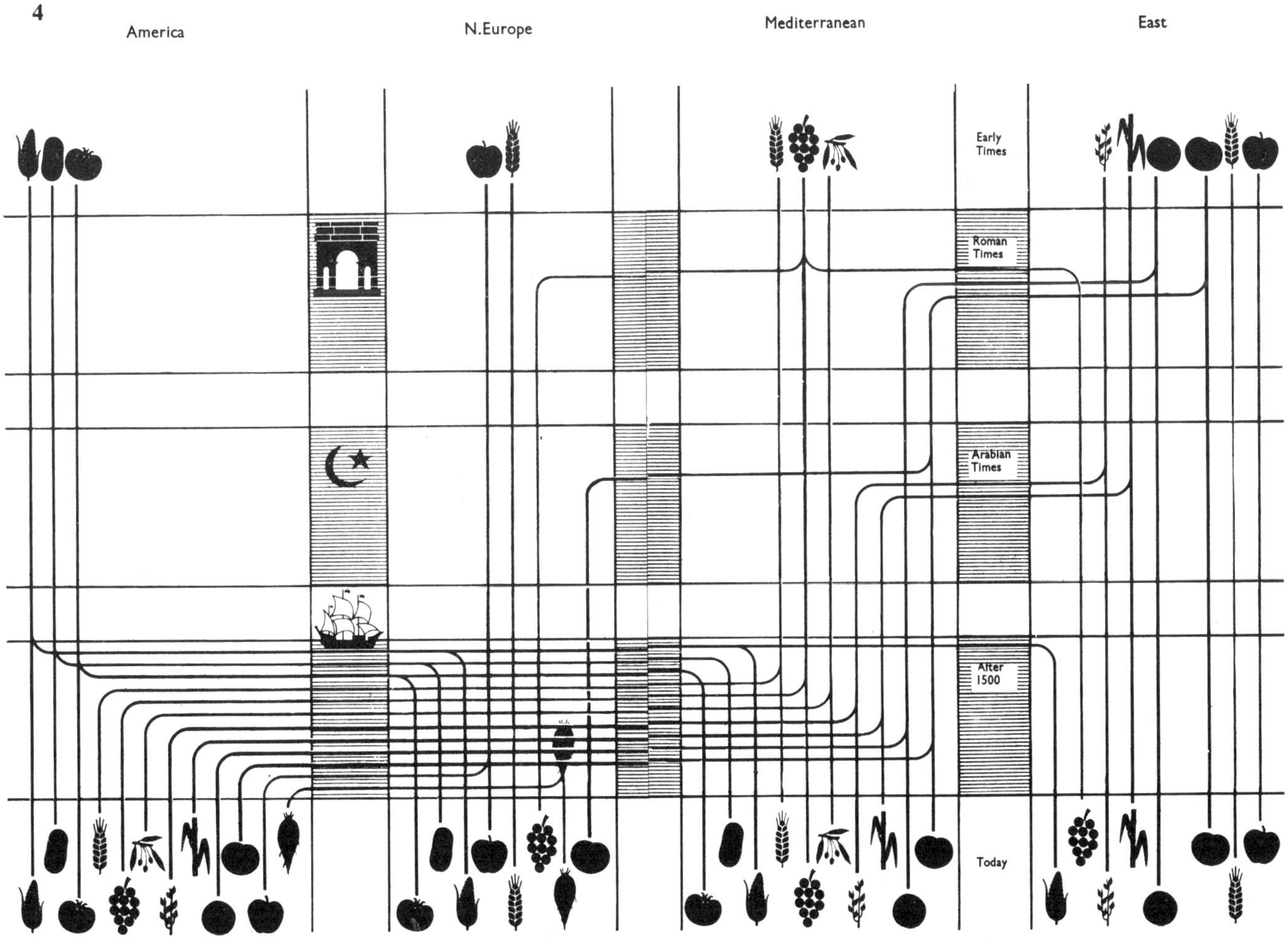

Time chart

1 Chart from a book, showing English history. Each distinct period of time is colour coded. The labels are executed in three ways: square to show definite birth dates and deaths and accurate representation of each lifetime on chart; rounded to show lifetimes not represented accurately on chart: scalloped to show dates only approximately known. Note the way the life *and* reign of kings are shown. H. A. Vetter, *English History*, Architectural Press

2 Diagram showing the year 1962 in circular form. This ingenious calendar demonstrates the simple idea of space-saving, through a more usual form. The year drawn as a straight line with exactly the same dots would appear to be approx 26″ long! Working days black, holidays and weekends red. Kast & Ehinger

On pp 136-37

1 Chart from an educational book, showing world history 1800–1970. One of four time charts in the book. The absolutely equal bands for each area allow only one symbol to be placed at each point in time. This results in a highly selective approach and only key items are shown. The use of a symbol (e.g. gun for World War I and bomb for World War II) enable easy comparison from region to region. Diagram by Aldus Books. *Il mondo della storia*, Mondadori

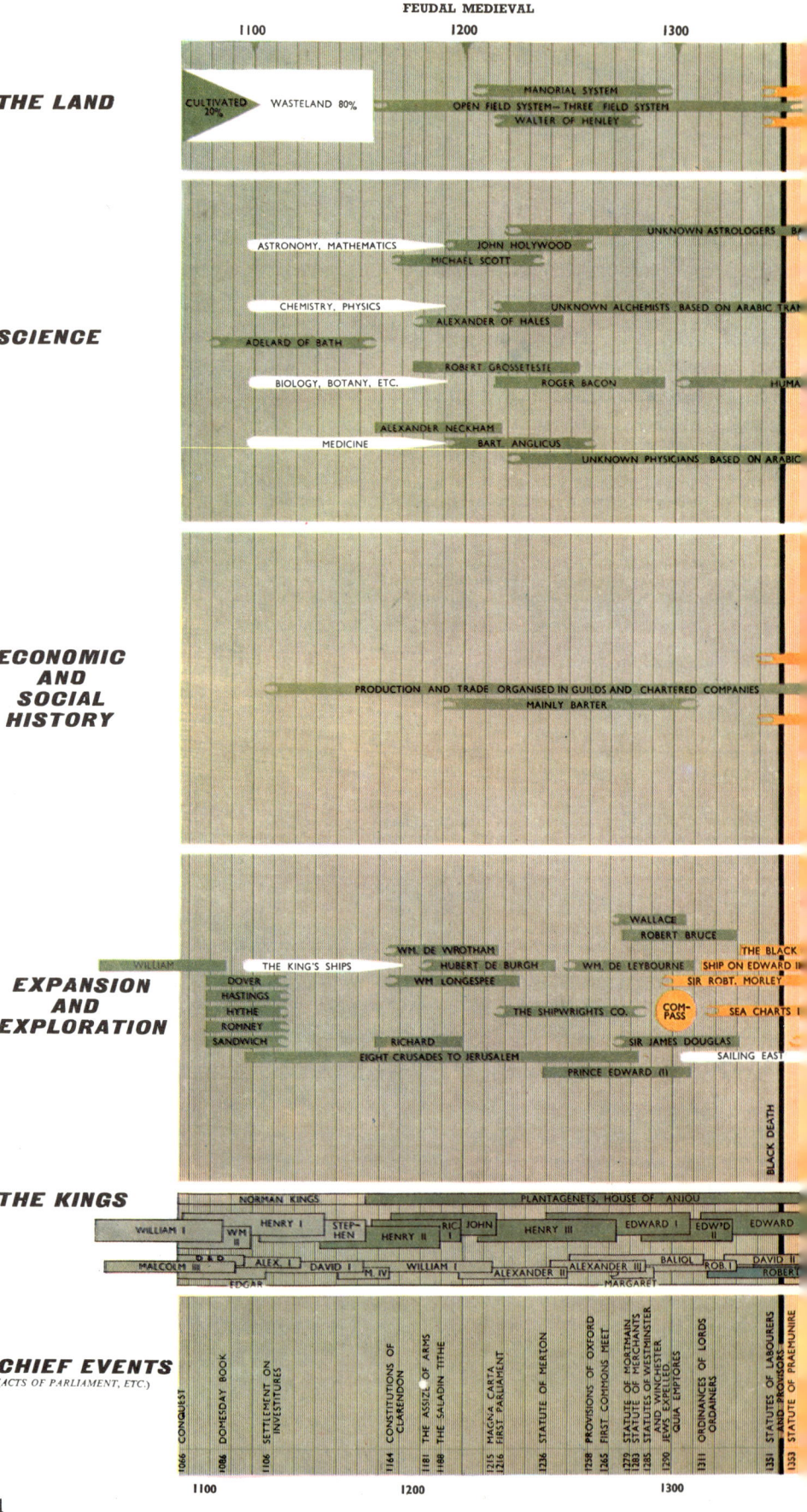

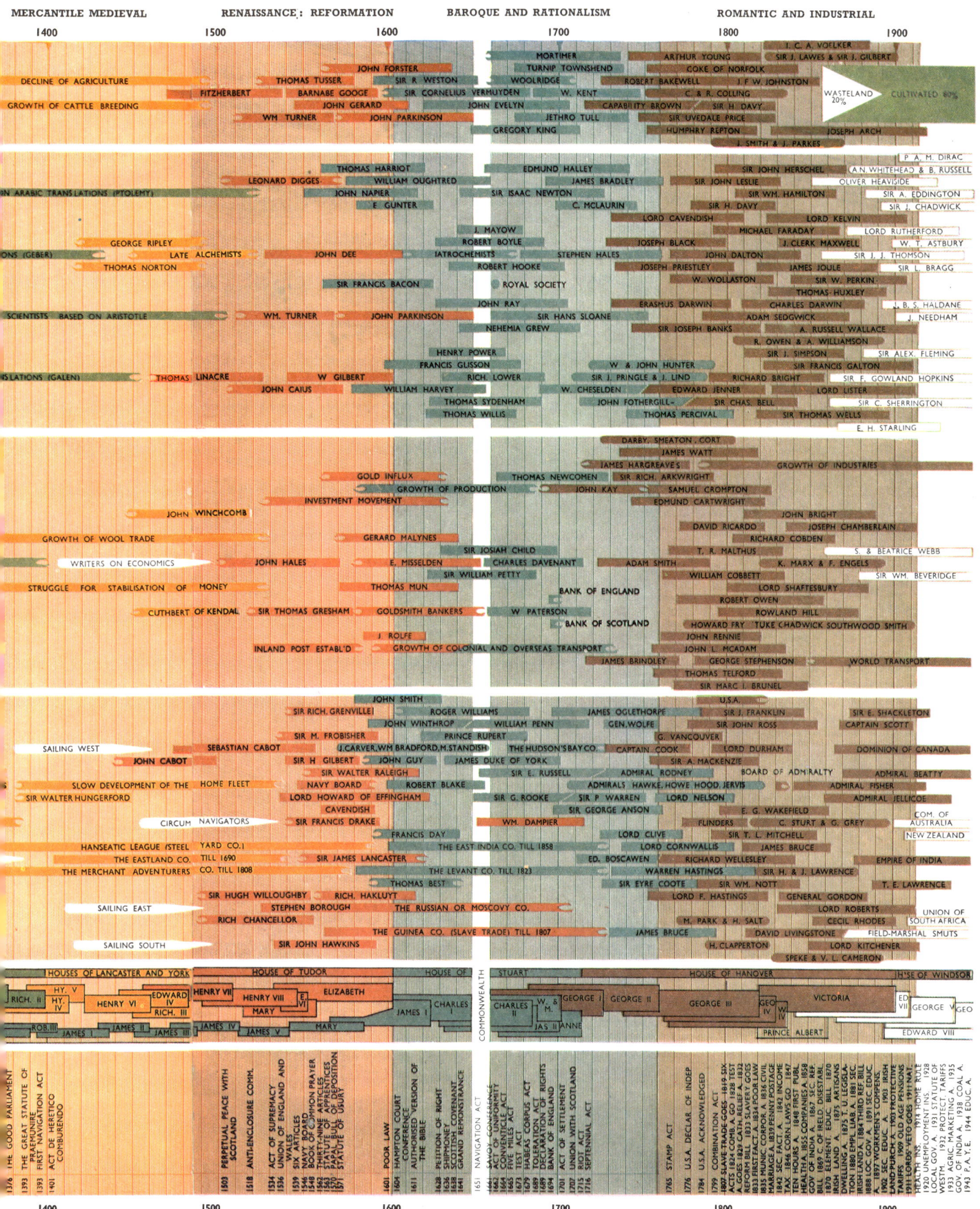

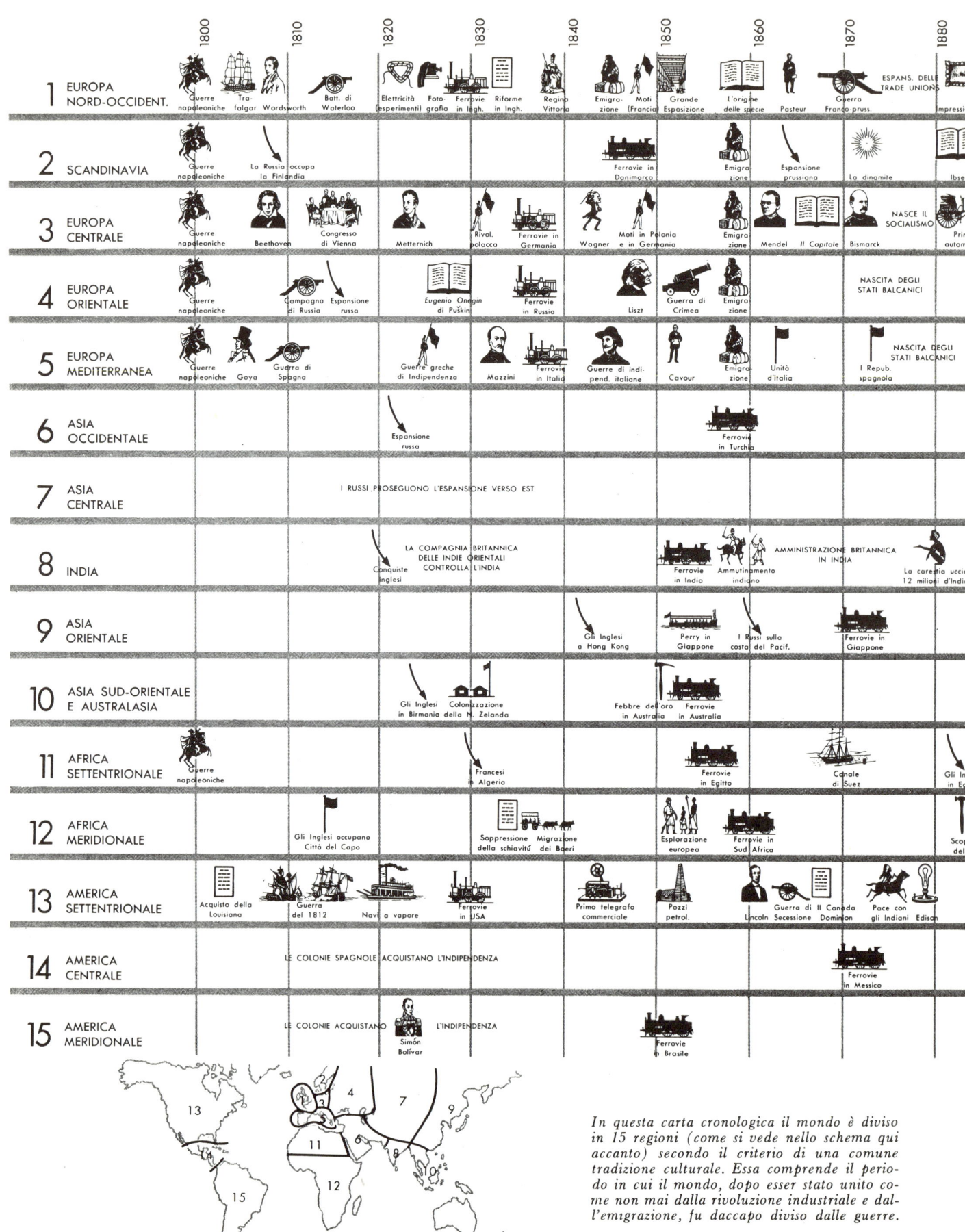

In questa carta cronologica il mondo è diviso in 15 regioni (come si vede nello schema qui accanto) secondo il criterio di una comune tradizione culturale. Essa comprende il periodo in cui il mondo, dopo esser stato unito come non mai dalla rivoluzione industriale e dall'emigrazione, fu daccapo diviso dalle guerre.

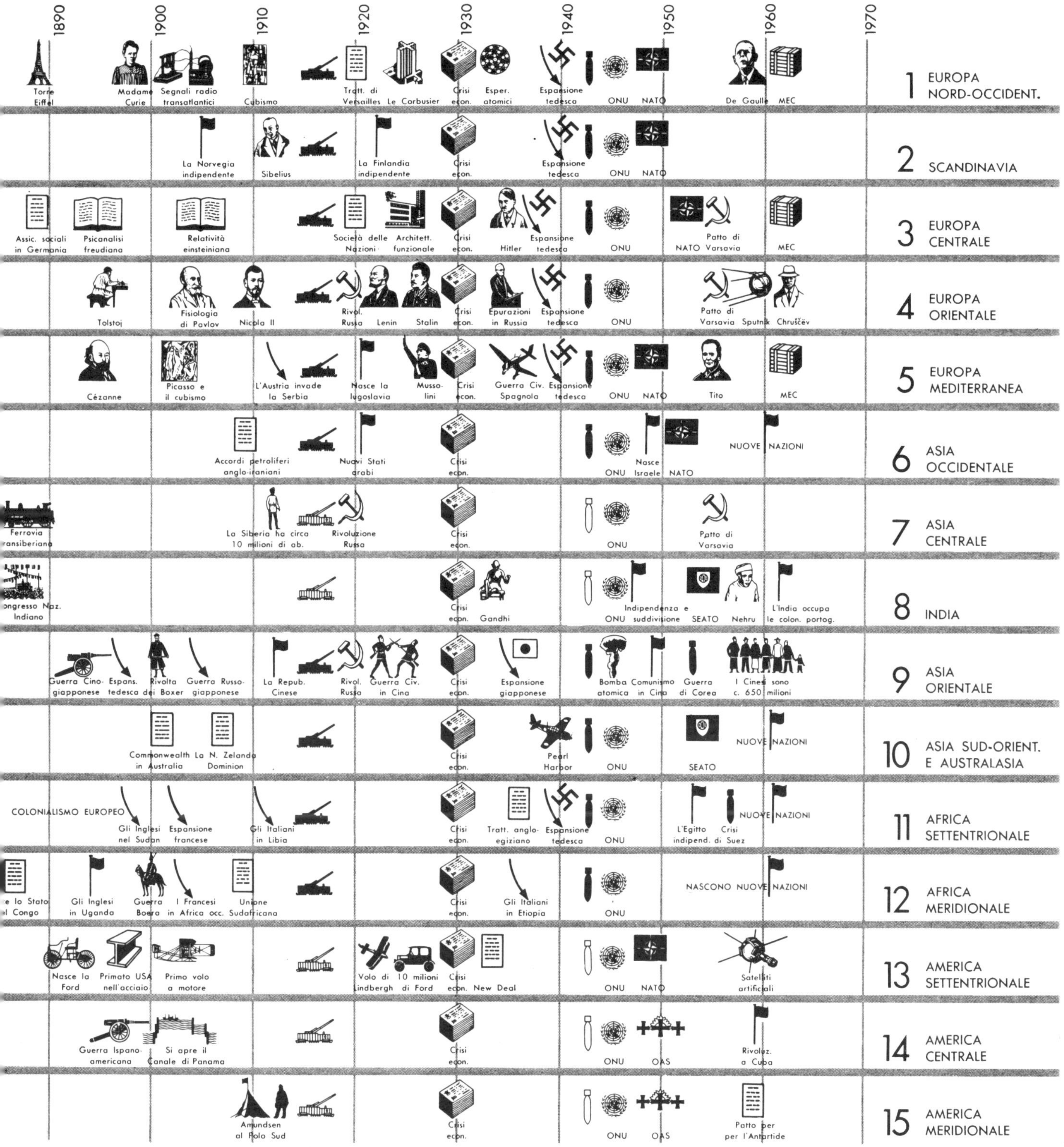

Il simbolo sta per "leggi, trattati, manifesti sia politici che sociali".

Opere e scritti di grande interesse.

Regioni in cui si combatté nella prima o seconda Guerra Mondiale.

Paesi intervenuti militarmente nella prima o seconda Guerra Mondiale.

Technique

Diagrams provide problems both of drawing and of design. The first is easier to solve than the second. Drawing can be made simpler by using the correct equipment and by learning to work in the right way. Finding this right way comes partly by practice and partly by hearing and reading about how diagrams can be drawn. Design of diagrams is more difficult and apart from personal ideas, and current graphic cliches, the most important factor is the original brief. Where is the diagram to be used, at what size, by what reproduction method, what information is it providing, is the reference source adequate, who will look at it?

The aim of this book has been to make the designer more aware of diagrams by looking at as wide a selection as possible, and by describing each method and its problems. The best conclusion to the book is to examine the problem of technique, to suggest ways of making diagrams easier to draw, and the design of better diagrams. By looking again at the 250 examples shown, comment in this section can be related to actual solutions and an assessment made of their appearance rather than their content. But first, drawing the diagram.

The base for many statistical diagrams is graph paper, obtainable with squares of $\frac{1}{4}''$, $\frac{1}{8}''$, $\frac{1}{10}''$, and 1mm. Special kinds of graph paper are available for certain work; semi-logarithmic (vertical scale only logarithmic), circular, percentage circular (circle divided into one hundred segments) and isometric, enabling three-dimensional figures to be constructed without a lot of angular measurement.

Besides the normal drawing equipment, it is advisable to have a ruling pen with a numbered dial which can be set for different thicknesses of line (an easy way of ensuring consistent lines for each diagram). Percentage protractor, proportional dividers, and an adjustable set-square will be useful. Ovals present problems of drawing but there exist many templates to help. Bevelled edges are always advisable for equipment that is used for lines drawn against the edge.

It is worth learning some of the short cuts to solving problems of drawing. The simple method of dividing any fixed length of line into any given number of units is shown in this example.

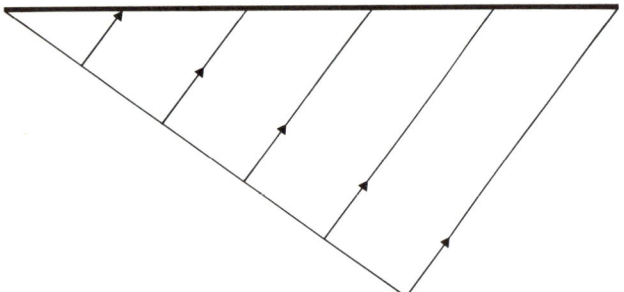

Books on geometry and engineering drawing will repay study and give information on such time-saving techniques.

In drawing diagrams, prime consideration must be given to consistency of line. Perhaps only one thickness of line will be used, but this should be exactly the same throughout the book. It goes without saying, therefore, that all diagrams should be drawn for the same size reduction, whether half-up or twice-up. If more lines are used, a definite coding system, showing thicknesses and their use, should be worked out and adhered to. The principle shown for engineering drawing applies equally to all diagrams, though they may have a different solution. A set of lines should be drawn on a separate board and keyed to the dial setting on the ruling pen described earlier.

The line used to indicate the part being labelled must also be consistent. There are various ways of labelling. This should be kept unobtrusive, avoiding the mannered approach which often takes the longest route.

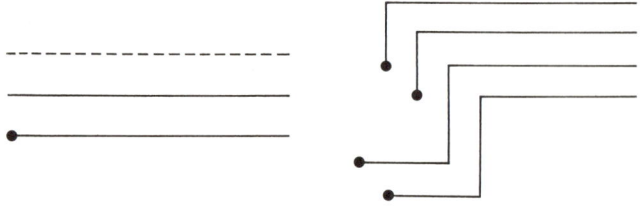

Arrow heads should also be standardized for a series of diagrams.

There are several easy ways of applying tone and texture. Firms such as Letraset and Zip-a-tone make a large range of tone and texture sheets which are easy to apply to artwork. Chart-Pak make textures and lines in tape form which can be used on involved line or bar graphs. These firms are continually increasing their ranges, and it is worth getting their latest catalogues.

Letraset also make sheets of symbols. The simple and more conventional are welcome. But at some point the designer evolves his own (such as the symbol for man page 56) and having drawn a master set gets photo prints to the sizes he will need. But in fields such as electrical diagrams, where symbols are repeated many times, any mechanical method of applying the symbol will obviously save time and give absolute consistency.

Consistency must apply to labelling. What kind of typeface should be used: sans serif, roman, hand drawn, capitals, lower case. Looking through the illustrations in this book suggests what the answer may be. The letter

138

form in keeping with the approach and line work of a diagram is sans serif. But look at a roman (page 18) and pen letter (pages 86 and 132); there might be a situation where they would be more appropriate. *Scientific American* always uses capital sans serif – is it easier to read than lower case?

The placing of labelling should be organized in the same way throughout and size should be consistent (some lettering may need to be larger for emphasis, but if it is too small it suggests that the diagram was given an excessive reduction.

The same careful approach should be followed in the placing and organization of the key, the position of the scale and title.

A three-dimensional approach should be considered for some diagram problems. This may involve making models specially (page 87) or using things readily available (page 141). The result has to be photographed sympathetically so that lighting and viewpoint emphasise the three-dimensional qualities – a poor photograph can make the result look hardly worth the trouble involved. One problem with this method is labelling: it is difficult to do well and is usually kept to the minimum. More often the impact of the diagram is of greater importance than the information, and one does not expect detailed statistics from this method.

An air brush technique will simulate three dimensions and enable more complicated labelling to be included.

In working out a statistical diagram from figures provided, it is important to rough out quickly two or three different ways of tackling the problem. Figures may be shown by various methods, but usually one will suggest itself as being right.

In working out an explanatory diagram from references, it is advisable, and sometimes essential, to get other versions of the same problem to compare the accuracy of information shown with the approach used, so that the new solution may produce a better diagram, which solves any inaccuracies or confusing elements.

Symbols may be needed for extensive use on diagrams and maps. There are various ways of repeating exactly the same symbol.
1 Photographic. A symbol drawn for a particular job. *La Terra in cui Viviamo*, Nicola Zanichelli Editore
2 Typeset. A series of symbols for use in guide books cut by the 'Monotype' Corporation Ltd
3 Letraset. Some of the electronic symbols available.

1

2   3

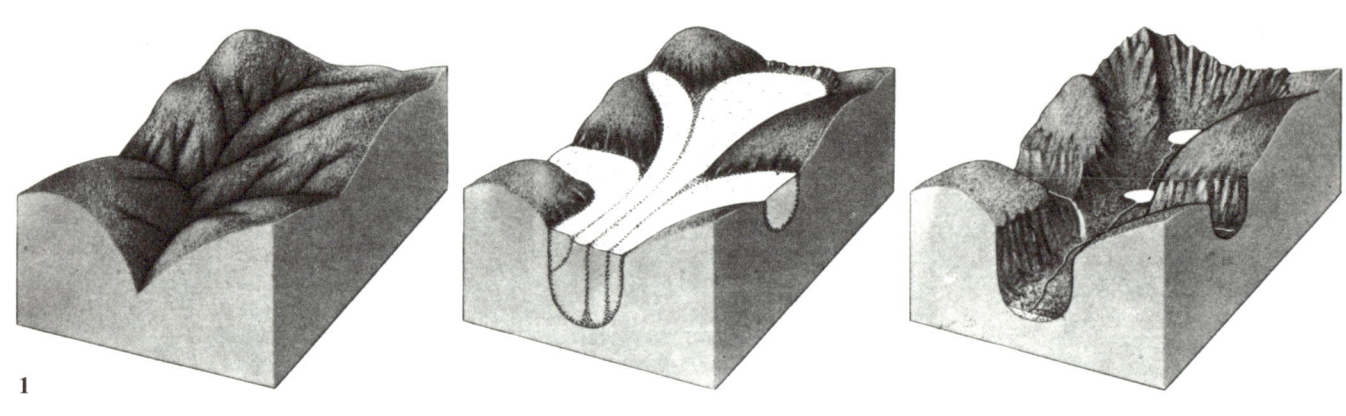

1

2

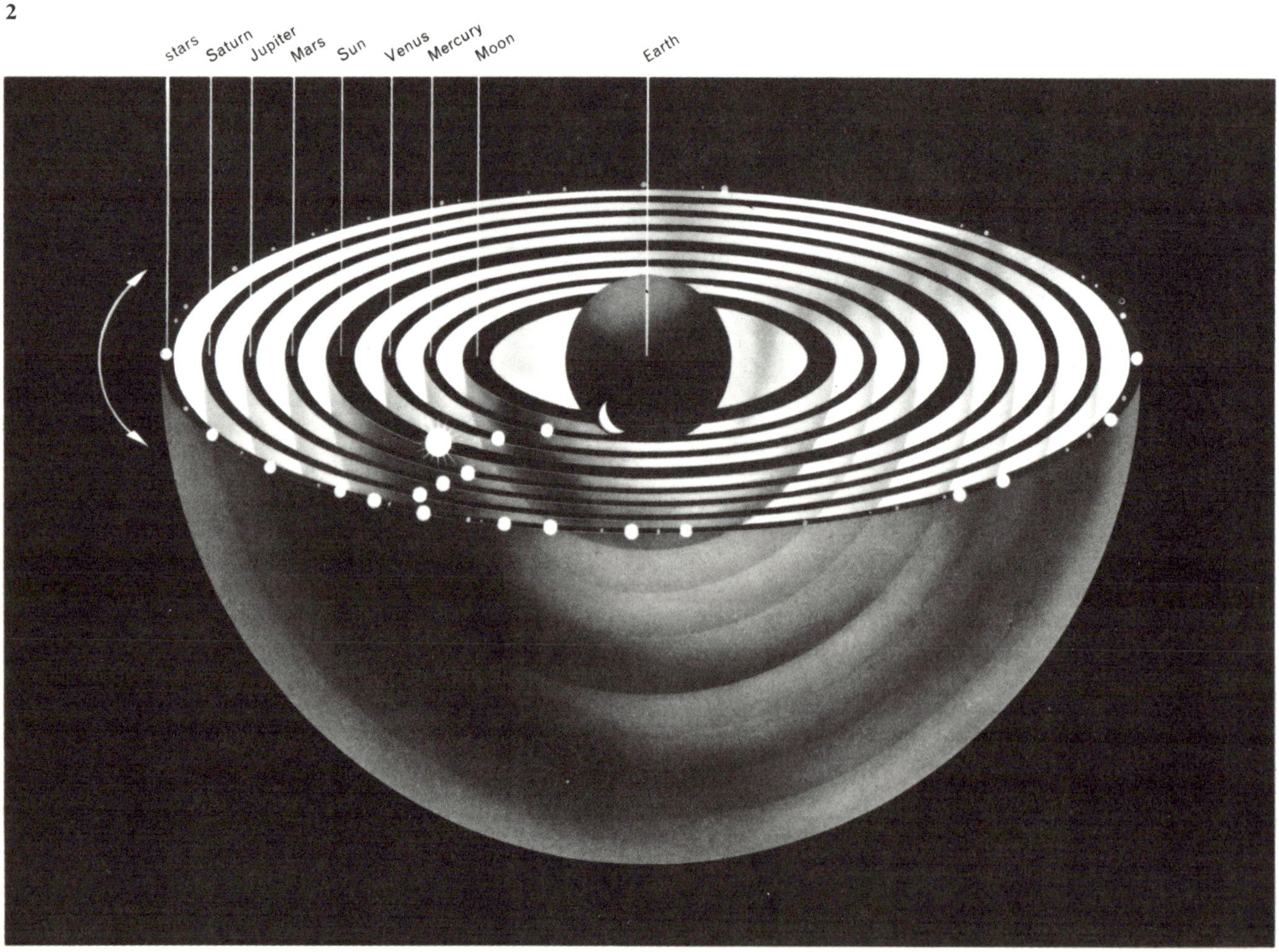

stars Saturn Jupiter Mars Sun Venus Mercury Moon   Earth

Three dimensions

1 Diagram from an educational book showing result of glacial action on the land-scape. Widely used by the geographers, this kind of diagram effectively illustrates features of a particular area; an added advantage is that geological information can be shown on the sides of the block.
W. E. Swinton, *The Story of Prehistoric Animals,* Rathbone Books

2 Diagram from a book showing early Greek astronomers' concept of the universe. Cut away sphere, air brush technique. Colin A. Ronan, *Man Probes the Universe,* Aldus Books

3 Part of a diagram from paperback encyclopedia showing astronomical co-ordinates. Three dimensions clearly shown by line shading. *dtv-Lexikon,* Deutscher Taschenbuch Verlag

4 Diagram from an advertisement showing proximities (considered 'near' together and confused, or 'far' apart and seldom confused) of ten vowel sounds. Air-brushed spheres in space. Bell Telephone Laboratories.

5 Schematic use of round headed pins on graph. Photo, Geoffrey Drury

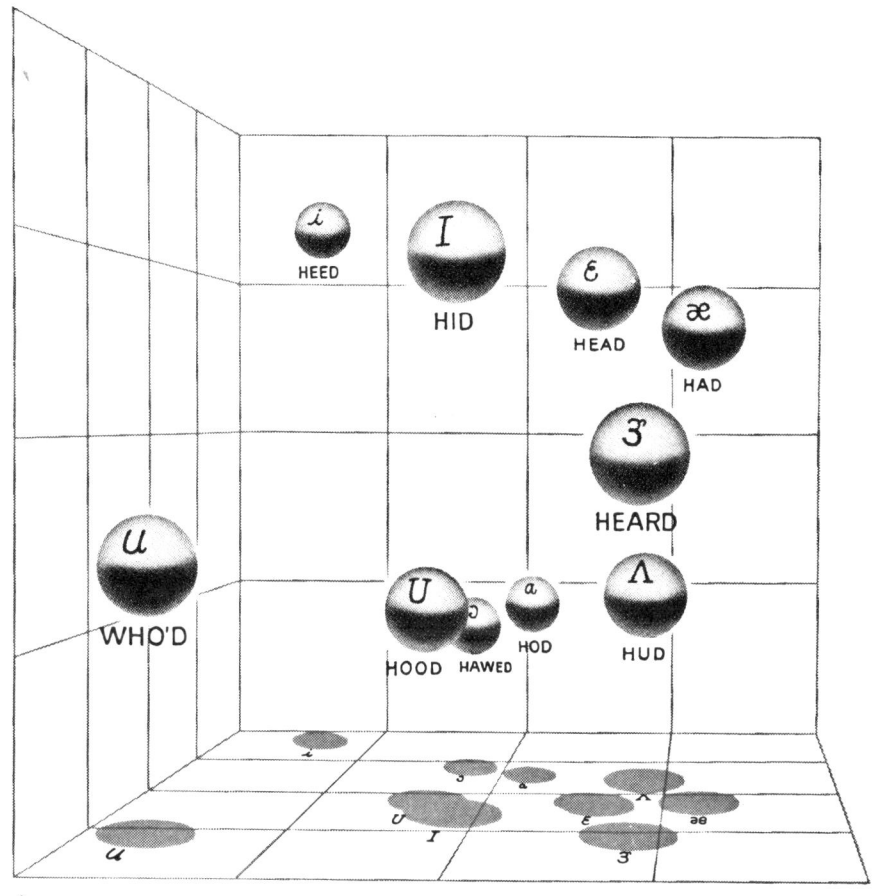
4

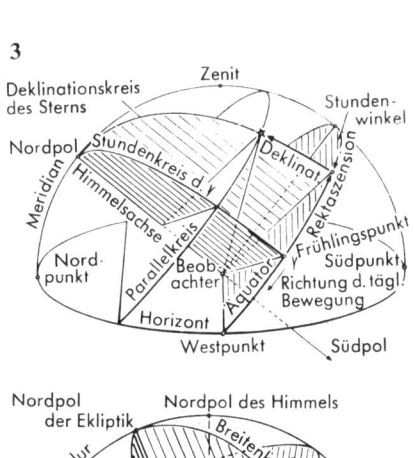
3

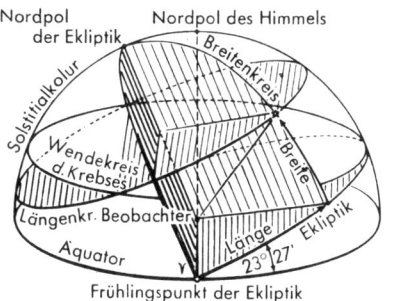

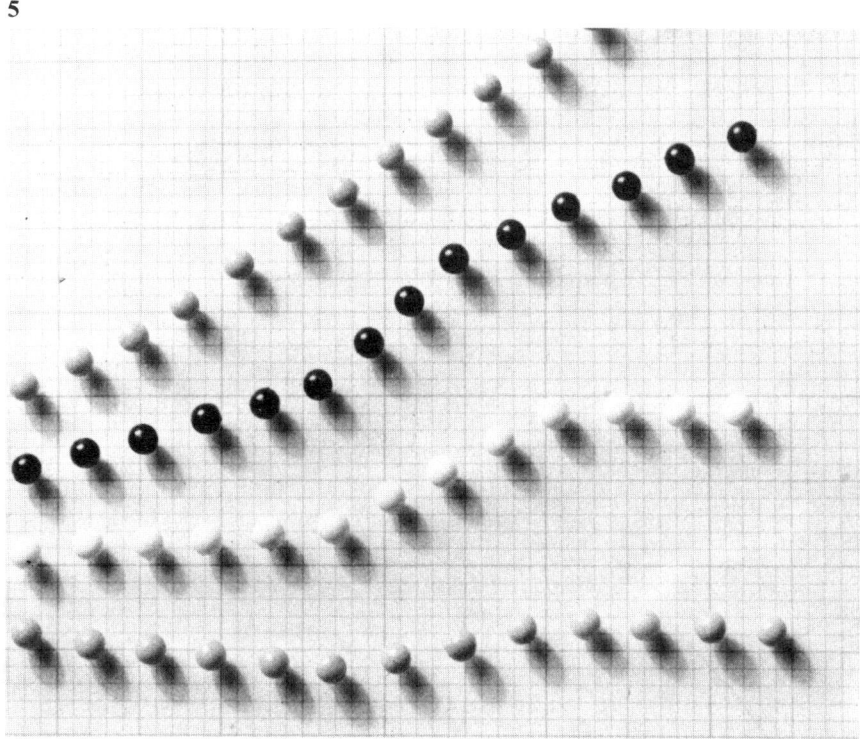
5

141

Engineering drawing

Type of line	Example	Application
continuous (thick)	A ————	visible outlines
continuous (thin)	B ————	dimension lines / projection or extension lines / hatching or sectioning / leader lines for notes / outlines of revolved sections
short dashes (thin)	C - - - - -	hidden details / portions to be removed
long chain (thin)	D — - — - —	centre lines / path lines for indicating movement / pitch circles
long chain (thick)	E — - — - —	cutting or viewing planes
short chain (thin)	F — - — - —	developed or false views / adjacent parts / feature located in front of a cutting plane / alternative position of moveable part
continuous wavy (thick)	G ∼∼∼∼	irregular boundary lines / short break lines
ruled line and short zig-zags	H —⋏⋏—⋏⋏—	long break lines

1

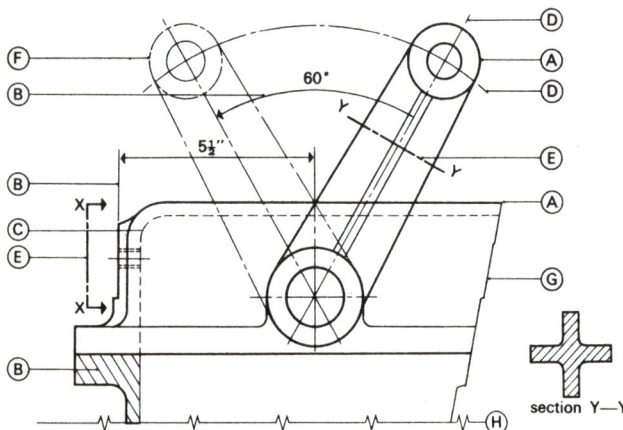

2

1 Types of lines for engineering drawings recommended in BS 308, a publication by the British Standards Institution, recommending the adoption of a number of conventions found generally acceptable in engineering. It takes into account practices recommended by the International Organisation for Standardisation.

2 Drawing using these recommended lines British Standards Institution/Penguin Education.

3 Isometric projection.

4 Axometric projection.

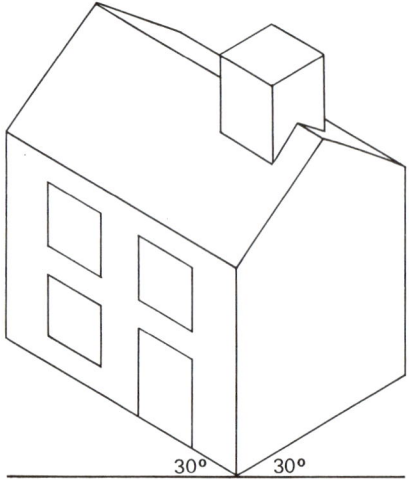

3

4

Bibliography

Since at the moment of writing there is no book published in England on diagrams for the graphic designer it is necessary to go to other fields for books to read.

G. C. Dickinson, *Statistical Mapping and the Presentation of Statistics,* Edward Arnold, introduces the subject to someone embarking on geography, economics or the social sciences. Avoiding mathematics, it is an excellent survey. There is a series of worked examples and particularly interesting are those giving the same information using different methods. There is a useful chapter with sources of statistics – where to get information to originate diagrams.

F. J. Monkhouse and H. R. Wilkinson, *Maps and Diagrams,* University Paperbacks, Methuen, is a reference book for the compilation and construction of maps and diagrams for the geographer; more specialist, but worth studying because of its comprehensive coverage of that field.

Arthur H. Robinson, *Elements of Cartography,* John Wiley. Problems of the map maker in detail.

Erwin Raisz, *Principles of Cartography,* McGraw-Hill.

T. W. Birch, *Maps, Topographical and Statistical,* Oxford University Press. A general survey of problems for the geographer, emphasis more on maps.

The Monotype Recorder, Precision in Map Making, volume 43, number 1, summer 1964. Interesting experiment by Cartographic Department of the Oxford University Press. Use of different type faces to label maps.

Signs and Symbols in Graphic Communication, Design Quarterly 62, Walker Art Center. Problems of the symbol investigated.

A. C. Parkinson, *General Engineering Drawing,* Pitman. A course in drawing for the engineer but useful hints on basic problems of drawing geometric and mechanical things.

I. H. Morris and J. C. Scott, *Geometrical Drawing,* Longmans. Solutions to every kind of problem in geometrical drawing.

Acknowledgments

First acknowledgment must be made to the numerous diagram artists whose work appears in this book. In so many cases they are not named by the publishers. And because of this lack of information I decided to acknowledge the source rather than credit individual diagram artists. But without them such a book would not be possible. I must thank all publishers and firms for giving me permission to reproduce diagrams from their books and publications.

Two publishers, Aldus Books and Penguin Books, have contributed indirectly to this book. As a designer working on their projects I gained experience and understanding of the subject and without this work my knowledge and use of diagrams would be much more limited.

Suggestions and help have also come from designers. I wish to thank particularly Bruce Robertson whose work for *History of the 20th Century* is illustrated here and provides thought-provoking solutions to the diagram problem. I have also benefited from Felix Gluck's experience of continental design and from his comments at layout stage.

Donald Holden of Watson-Guptill made helpful suggestions at the start of the book and his interest and enthusiasm concentrated my effort on the main problems of the subject. He helped with American sources.

I must also thank my publishers for their patience (the time-consuming compilation of this book has sometimes had to take second place to work with more demanding deadlines) and in particular to Robin Wright for sorting out the exact meaning of a text written by someone more capable of designing than writing.

In a work of this nature, which is necessarily dependent upon illustrations reproduced from a multitude of sources from all over the world, it is often difficult and sometimes impossible to ascertain whether or not a particular illustration is in copyright, and, if it is, who is the owner. The author and publishers sincerely apologise if they have unwittingly infringed copyright in any illustration reproduced in this volume.

Index

The index refers to the subject of the diagram or map. Since certain diagrams have similar or related subjects the index is arranged in groups with numerous sub-headings. In this way it is possible to compare different diagram methods being used for similar problems.

aviation
   aircraft passengers 30
   aircraft performance 64-5
   airline routes 103
   airport to city journeys 51
   flights to Canada 63
   flights over the Atlantic 103
   speeds 12

biology
   classification of flora and fauna 120
   evolution 127
   height of a boy 10
   inheritance factor 124
   insect (distance travelled) 41
   life cycle of a beetle 130
   marine life (map) 95
   metabolism 22
   physique 23
   plant distribution (map) 88
   speed of man/animals 63

energy
   balance for Switzerland 110
   consumption past/future 21
   from sun 113
   source 127
   US annual consumption 17
   use v. gross national production 25

finance
   cost of a bar of chocolate 47
   cost of a book 54
   prices of materials 32
   sales/price of chocolate 38
   share price range 34
   share values 21
   sources/use of funds 47
   trading profit 18
   UK currency reserves 10
   UK wealth/income 58
   US investment (map) 105

food
   fish (map) 92
   home produced/imported 61
   meals at restaurants 53
   origins of fruit and vegetables 133
   wheat (map) 85, 88, 93, 104

geography/geology
   air pressure (map) 98
   composition of earth's crust 52
   earth's land area 49
   glacial action 140
   rainfall (map) 80
   size of continents 43
   temperature 12, 50, 51
   tides 34
   use of land 46
   vegetation (map) 81
   weather (map) 69
   wind rose 51

industry
   home and factory weaving 57
   manufacturing processes 109, 111, 112, 114, 115
   occupations of labour force 19
   oil 13, 35, 44, 106-7
   organization of General Motors 121
   parts of a company 45
   raw materials 54
   sales of chocolate 48
   unemployment 11, 33
   use of working hours 50
   'who-produces-what' 61

motor car
   acceleration/braking 41
   energy flow 110
   engine 116
   motor oil 40
   production 31
   road deaths 27
   traffic flow (map) 104

population
   according to area and race 59
   Africa (map) 93
   animal 28, 43
   density (map) 81, 90, 95
   density profile 16
   distorted maps 98-101
   distribution in London 85
   displacement as result of Second World War (map) 89
   New York 17
   pyramids 36-7
   shift from rural 60
   United States 13
   urban 35, 49
   world 9, 19, 55, 57

politics
   balance of power 1914 (map) 87
   'Cold War' 14
   electoral constituencies (map) 101
   empires 42, 132
   French election (map) 97
   French party representation (map) 91
   NATO network (map) 94
   regions of USSR (map) 98
   spheres of influence 24
   UK party support 11
   United Nations Organization 119
   US government 119

production
   aluminium 52
   car 31
   fibres 20
   industrial 33
   iron/steel 17
   metals (map) 96
   office machinery 54
   oil 13, 35, 106-7
   pig iron 27
   raw materials 54
   vegetable oils 29
   wine 16

religion
   in Asia (map) 95

   Society of Friends meeting houses (map) 88
   spheres of influence (map) 92
   spread of Islam (map) 80

science/technology
   aircraft performance 64-5
   astronomy 140-1
   aviation speeds 12
   distribution of scientists (map) 99
   fibre stress-strain 11
   fibre weathering resistance 48
   generating electricity 15
   hydro-electric power potential 29
   nuclear power/fossil power 39
   plasma containment 22
   power station use 18
   research programmes 128
   sound 141
   thermal generating capacity 29
   turbines 117

society
   age for military service, marriage, etc. 128
   education 40, 121, 133
   houses (type and rent) 62
   immigration (map) 84
   input/output transactions of a city 44
   languages (map) 79, 126
   medieval structure 118
   sport 14, 76
   university libraries (map) 86
   use of land 46
   water consumption of New York 17

time
   calendar 134
   distance in time from London (map) 100
   increasing leisure 53
   radio programmes 130
   to get a meal 53

trade
   duty on imports 30
   exports (map) 96
   freight handled by European ports (map) 86
   French trade 45
   harbours (annual turnover) 84
   home-produced/imported food 61
   import/export to EFTA 32
   oil trade 106/7
   trading company activities 115
   wheat trade (map) 93, 104

transport/travel
   between two towns 60
   mileage 128
   railway freight (map) 105
   ships crossing Atlantic 103
   time-distance from London (map) 100
   traffic flow 104
   Underground train timetable 15

war
   battles 77
   expenditure on First World War 55
   submarines sunk 62
   troops in Vietnam 27